# GUIDELINES FOR
# MECHANICAL INTEGRITY SYSTEMS

This book is one of a series of publications available from the Center for Chemical Process Safety. A computer list of titles appears at the end of this book.

# GUIDELINES FOR MECHANICAL INTEGRITY SYSTEMS

Center for Chemical Process Safety of the
American Institute of Chemical Engineers

An AIChE Industry
Technology Alliance

A JOHN WILEY & SONS, INC., PUBLICATION

Copyright © 2006 by American Institute of Chemical Engineers. All rights reserved.

A joint publication of the Center for Chemical Process Safety of the American Institute of Chemical Engineers and John Wiley & Sons, Inc.

Published by John Wiley & Sons, Inc., Hoboken, New Jersey.
Published simultaneously in Canada.

No part of this publication may be reproduced, stored in a retrieval system, or transmitted in any form or by any means, electronic, mechanical, photocopying, recording, scanning, or otherwise, except as permitted under Section 107 or 108 of the 1976 United States Copyright Act, without either the prior written permission of the Publisher, or authorization through payment of the appropriate per-copy fee to the Copyright Clearance Center, Inc., 222 Rosewood Drive, Danvers, MA 01923, (978) 750-8400, fax (978) 750-4470, or on the web at www.copyright.com. Requests to the Publisher for permission should be addressed to the Permissions Department, John Wiley & Sons, Inc., 111 River Street, Hoboken, NJ 07030, (201) 748-6011, fax (201) 748-6008, or online at http://www.wiley.com/go/permission.

Limit of Liability/Disclaimer of Warranty: While the publisher and author have used their best efforts in preparing this book, they make no representations or warranties with respect to the accuracy or completeness of the contents of this book and specifically disclaim any implied warranties of merchantability or fitness for a particular purpose. No warranty may be created or extended by sales representatives or written sales materials. The advice and strategies contained herein may not be suitable for your situation. You should consult with a professional where appropriate. Neither the publisher nor author shall be liable for any loss of profit or any other commercial damages, including but not limited to special, incidental, consequential, or other damages.

For general information on our other products and services or for technical support, please contact our Customer Care Department within the United States at (800) 762-2974, outside the United States at (317) 572-3993 or fax (317) 572-4002.

Wiley also publishes its books in a variety of electronic formats. Some content that appears in print may not be available in electronic format. For information about Wiley products, visit our web site at www.wiley.com.

*Library of Congress Cataloging-in-Publication Data:*

Guidelines for mechanical integrity systems / Center for Chemical Process Safety (CCPS).
    p. cm.
    Includes index.
    ISBN-13: 978-0-8169-0952-0 (cloth/cd)
    ISBN-10: 0-8169-0952-0 (cloth/cd)
    1. Chemical equipment—Maintenance and repair.  2. Chemical plants—Safety measures.  3. Plant maintenance.  I. American Institute of Chemical Engineers. Center for Chemical Process Safety.
    TS191.G85 2006
    660'.2830288—dc22                                                    2006008220

Printed in the United States of America.

10 9 8 7 6 5 4 3 2 1

It is sincerely hoped that the information presented in this document will lead to an even more impressive safety record for the entire industry; however, neither the American Institute of Chemical Engineers, its consultants, CCPS Technical Steering Committee and Subcommittee members, their employers, their employers officers and directors, nor ABSG Consulting and its employees warrant or represent, expressly or by implication, the correctness or accuracy of the content of the information presented in this document. As between (1) American Institute of Chemical Engineers, its consultants, CCPS Technical Steering Committee and Subcommittee members, their employers, their employers officers and directors, and ABSG Consulting, and its employees, and (2) the user of this document, the user accepts any legal liability or responsibility whatsoever for the consequence of its use or misuse.

# CONTENTS

| | |
|---|---|
| List of Tables | xiii |
| List of Figures | xv |
| Items on the CD Accompanying This Book | xvii |
| Acronyms and abbreviations | xxi |
| Glossary | xxv |
| Acknowledgments | xxix |
| Preface | xxxi |
| Management Overview of the Guidelines | xxxiii |

## 1 INTRODUCTION 1

| | | |
|---|---|---|
| 1.1 | What is Mechanical Integrity? | 2 |
| 1.2 | Relationship to Other Programs | 3 |
| 1.3 | Expectations for the MI Program | 3 |
| 1.4 | The Effect of RAGAGEPs | 5 |
| 1.5 | Structure of this Guidelines Book | 6 |
| 1.6 | References | 8 |

## 2 MANAGEMENT RESPONSIBILITY 9

| | | |
|---|---|---|
| 2.1 | Facility Leadership's Roles and Responsibilities | 9 |
| 2.1.1 | Organizational Roles and Responsibilities | 10 |
| 2.1.2 | Roles and Responsibilities Matrix | 10 |
| 2.1.3 | Reporting Mechanisms | 11 |
| 2.1.4 | Auditing | 14 |
| 2.2 | Technical Assurance Responsibilities | 14 |
| 2.2.1 | Defining Acceptance Criteria | 14 |
| 2.2.2 | Providing Technical Content | 15 |
| 2.2.3 | Establishing Metrics | 15 |
| 2.2.4 | Ensuring Technical Review | 16 |

## 3 EQUIPMENT SELECTION 17

3.1 Reviewing Program Objectives 17
3.2 Establishing Equipment Selection Criteria 18
3.3 Defining Level of Detail 21
3.4 Documenting the Equipment Selection 22
3.5 Equipment Selection Roles and Responsibilities 23
Apendix 3A. Sample Guidelines for Selecting Equipment for the MI Program 25

## 4 INSPECTION, TESTING, AND PREVENTIVE MAINTENANCE 29

4.1 ITPM Task Planning 30
    4.1.1 ITPM Task Selection 30
    4.1.2 Developing Sampling Criteria 42
    4.1.3 Other ITPM Task Planning Considerations 44
    4.1.4 ITPM Task Scheduling 45
4.2 Task Execution and Monitoring 46
    4.2.1 Defining Acceptance Criteria 46
    4.2.2 Equipment and ITPM Task Results Documentation 47
    4.2.3 ITPM Task Implementation and Execution 49
    4.2.4 ITPM Task Results Management 49
    4.2.5 Task Schedule Management 51
    4.2.6 ITPM Program Monitoring 52
4.3 ITPM Program Roles and Responsibilities 53
4.4 References 53
Appendix 4A. Common Predictive Maintenance and Nondestructive Testing Techniqes 58
Appendix 5A. Sample Training Survey 75

## 5 MI TRAINING PROGRAM 61

5.1 Skills/Knowledge Assessment 61
5.2 Training For New and Current Workers 64
5.3 Verification and Documentation of Training Effectiveness 64
5.4 Certification 66
5.5 Ongoing and Refresher Training 69
5.6 Training for Technical Personnel 69
5.7 Contractor Issues 71

| | 5.8 | Roles and Responsibilities | 71 |
|---|---|---|---|
| | 5.9 | References | 74 |

# 6 MI PROGRAM PROCEDURES 77

| | | |
|---|---|---|
| 6.1 | Types of Procedures Supporting the MI Program | 79 |
| 6.2 | Identification of MI Procedure Needs | 81 |
| 6.3 | Procedure Development Process | 83 |
| 6.4 | MI Procedure Format and Content | 87 |
| 6.5 | Other Sources of MI Procedures | 90 |
| 6.6 | Implementing and Maintaining MI Procedures | 91 |
| 6.7 | Procedure Program Roles and Responsibilities | 92 |
| 6.8 | References | 92 |

# 7 QUALITY ASSURANCE 95

| | | |
|---|---|---|
| 7.1 | Design | 96 |
| 7.2 | Procurement | 97 |
| 7.3 | Fabrication | 98 |
| 7.4 | Receiving | 99 |
| 7.5 | Storage and Retrieval | 99 |
| 7.6 | Construction and Installation | 100 |
| 7.7 | In-service Repairs, Alterations, and Rerating | 101 |
| 7.8 | Temporary Installations and Temporary Repairs | 102 |
| 7.9 | Decommissioning/Reuse | 103 |
| 7.10 | Used Equipment | 103 |
| 7.11 | Spare Parts | 104 |
| 7.12 | Contractor-Supplied Equipment and Materials | 104 |
| 7.13 | QA Program Roles and Responsibilities | 104 |
| 7.14 | References | 104 |
| Appendix 7A. Design Review Suggestions | | 107 |
| Appendix 7B. Sample Vendor QA Plan | | 110 |
| Appendix 7C. Positive Material Identification | | 112 |
| Appendix 7D. Sample Service QA Plan | | 116 |
| Appendix 8A. Fitness for Service (FS) | | 131 |

# 8 EQUIPMENT DEFICIENCY MANAGEMENT 119

| | | |
|---|---|---|
| 8.1 | Equipment Deficiency Management Process | 119 |
| 8.2 | Acceptance Criteria | 120 |
| 8.3 | Equipment Deficiency Identification | 122 |
| 8.4 | Responding to Equipment Deficiencies | 123 |

|  |  |  |
|---|---|---|
| 8.5 | Equipment Deficiency Communication | 127 |
| 8.6 | Permanent Correction of Equipment Deficiencies | 127 |
| 8.7 | Deficiency Management Roles and Responsibilities | 128 |
| 8.8 | Reference | 128 |

# 9 EQUIPMENT-SPECIFIC INTEGRITY MANAGEMENT 135

|  |  |  |  |
|---|---|---|---|
| 9.1 | Fixed Equipment | | 137 |
| 9.2 | Relief and Vent Systems | | 142 |
| 9.3 | Instrumentation and Controls | | 144 |
| 9.4 | Rotating Equipment | | 145 |
| 9.5 | Fired Equipment | | 151 |
| 9.6 | Electrical Systems | | 151 |
| 9.7 | Fire Protection Systems | | 153 |
| 9.8 | Miscellaneous Equipment | | 154 |
|  | 9.8.1 | Ventilation and Purge Systems | 154 |
|  | 9.8.2 | Protective Systems | 155 |
|  | 9.8.3 | Solids-handling Systems | 156 |
|  | 9.8.4 | Safety-critical Utilities | 157 |
|  | 9.8.5 | Other Safety Equipment | 157 |
| 9.9 | Equipment-specific MI Activity Matrices | | 158 |
| 9.10 | References | | 182 |

# 10 MI PROGRAM IMPLEMENTATION 183

|  |  |  |  |
|---|---|---|---|
| 10.1 | Budgeting and Resources | | 183 |
|  | 10.1.1 | Program Development Resources | 183 |
|  | 10.1.2 | Initial Implementation Resources | 187 |
|  | 10.1.3 | Ongoing Efforts | 193 |
| 10.2 | Use of Software in MI Programs | | 196 |
|  | 10.2.1 | Use of CMMS | 197 |
|  | 10.2.2 | Other Software Used in MI Programs | 198 |
| 10.3 | Return on Investment (ROI) | | 200 |
|  | 10.3.1 | Improved Equipment Reliability | 200 |
|  | 10.3.2 | Cost Avoidance | 201 |
|  | 10.3.3 | Regulatory Compliance and Industry Association Commitments | 202 |
|  | 10.3.4 | Reduced Liability and Reduced Damage to Corporate Reputation | 202 |
| 10.4 | References | | 202 |

# 11 RISK MANAGMENT TOOLS 203

| | | |
|---|---|---|
| 11.1 | Introduction to Common Risk-based Analytical Techniques Used in MI Programs | 205 |
| 11.2 | Incorporating Risk into MI Decisions | 210 |
| 11.3 | FMEA/FMECA | 212 |
| 11.4 | RCM | 213 |
| 11.5 | Risk-based Inspection | 218 |
| | 11.5.1 Equipment and Process Data | 221 |
| | 11.5.2 Risk Modeling | 221 |
| | 11.5.3 Inspection Planning Strategies/Guidelines | 222 |
| | 11.5.4 Other RBI Program Issues | 223 |
| 11.6 | Protection Layer Analysis Techniques | 225 |
| 11.7 | References | 229 |

# 12 CONTINUOUS IMPROVEMENT OF MI PROGRAMS 231

| | | |
|---|---|---|
| 12.2 | Program Audits | 233 |
| 12.3 | Performance Measurement and Monitoring | 238 |
| 12.4 | Equipment Failure and Root Cause Analyses | 240 |
| | 12.4.1 Failure Analysis | 243 |
| | 12.4.2 Root Cause Analysis | 245 |
| 12.5 | References | 248 |

# INDEX 249

# LIST OF TABLES

| | | |
|---|---|---|
| TABLE 1-1 | Potential MI Interfaces with Other Programs | 4 |
| TABLE 1-2 | Chapters Addressing Management Systems for MI Activities | 7 |
| TABLE 2-1 | Roles and Responsibilities Matrix for MI Program Management | 12 |
| TABLE 3-1 | Example Roles and Responsibilities Matrix for Equipment Selection | 24 |
| TABLE 4-1 | Typical Equipment File Information for Selected Types of Equipment | 34 |
| TABLE 4-2 | Example ITPM Task Selection Decision Matrix (Ref. 4-1) | 39 |
| TABLE 4-3 | Factors Affecting ITPM Tasks for Relief Valves, Instrumentation, and Rotating Equipment | 41 |
| TABLE 4-4 | Example ITPM Plan in Tabular Format | 43 |
| TABLE 4-5 | Example Roles and Responsibilities Matrix for the ITPM Task Planning Phase | 55 |
| TABLE 4-6 | Example Roles and Responsibilities Matrix for the ITPM Task Execution and Monitoring Phase | 56 |
| TABLE 5-1 | Training Approach Considerations | 65 |
| TABLE 5-2 | General Electrician Training Matrix | 67 |
| TABLE 5-3 | Widely Accepted MI Certifications | 68 |
| TABLE 5-4 | Mechanical Engineer Training Plan | 70 |
| TABLE 5-5 | Example Roles and Responsibilities Matrix for the MI Training Program | 72 |
| TABLE 6-1 | Example MI Procedures | 80 |
| TABLE 6-2 | Example Risk Ranking Results for Procedure Determination | 84 |
| TABLE 6-3 | Example Roles and Responsibilities Matrix for the MI Procedure Program | 93 |
| TABLE 7-1 | Typical Design Code Applications | 96 |
| TABLE 7-2 | Sample of Codes and Standards Having QA Requirements Applicable to Repair, Alteration, and Rerating | 101 |

| | | |
|---|---|---|
| TABLE 7-3 | Example Roles and Responsibilities Matrix for the QA Program | 106 |
| TABLE 8-1 | Acceptance Criteria Resources | 121 |
| TABLE 8-2 | Examples of Acceptance Criteria for Common Types of Equipment | 124 |
| TABLE 8-3 | Example Roles and Responsibilities Matrix for Equipment Deficiency Resolution | 129 |
| TABLE 9-1 | RAGAGEPs for Pressure Vessels | 138 |
| TABLE 9-2 | RAGAGEPs for Atmospheric and Low-pressure Storage Tanks | 139 |
| TABLE 9-3 | RAGAGEPs for Process Piping | 140 |
| TABLE 9-4 | RAGAGEPs for Pressure Relieving Devices | 143 |
| TABLE 9-5 | RAGAGEPs for Instrumentation and Controls | 146 |
| TABLE 9-6 | RAGAGEPs for Pumps | 147 |
| TABLE 9-7 | RAGAGEPs for Compressors | 148 |
| TABLE 9-8 | RAGAGEPs for Turbines | 149 |
| TABLE 9-9 | RAGAGEPs for Fans and Gearboxes | 150 |
| TABLE 9-10 | RAGAGEPs for Fired Heaters and Furnaces | 152 |
| TABLE 9-11 | Summary of Commonly Used NFPA Codes for Fire Protection Systems | 153 |
| TABLE 9-12 | Summary of RAGAGEPs for Selected Safety Equipment | 158 |
| TABLE 9-13 | Mechanical Integrity Activities for Pressure Vessels | 159 |
| TABLE 9-14 | Mechanical Integrity Activities for Piping Systems | 162 |
| TABLE 9-15 | Mechanical Integrity Activities for Pressure Relief Valves | 165 |
| TABLE 9-16 | Mechanical Integrity Activities for SISs and ESDs | 168 |
| TABLE 9-17 | Mechanical Integrity Activities for Pumps | 172 |
| TABLE 9-18 | Mechanical Integrity Activities for Fired Heaters/Furnaces/Boilers | 176 |
| TABLE 9-19 | Mechanical Integrity Activities for Switch Gear | 179 |
| TABLE 10-1 | Summary of Resources Required for MI Program Development Activities | 185 |
| TABLE 10-2 | Typical Initial Implementation Tasks by Activity | 188 |
| TABLE 10-3 | Examples of Ongoing QA Activities | 195 |
| TABLE 11-1 | Summary of Risk-based Analytical Techniques | 206 |
| TABLE 11-2 | Sample FMEA Worksheet | 214 |
| TABLE 11-3 | Sample RCM FMEA Worksheet | 216 |
| TABLE 11-4 | Final Failure Management Strategy Selection for Pump 1A | 218 |
| TABLE 11-5 | Example Inspection Strategy for Pressure Vessel - External Deterioration Identified by External Visual Inspection | 223 |
| TABLE 11-6 | ISA S84.01 SILs | 228 |
| TABLE 12-1 | MI Audit Approach | 235 |

# LIST OF FIGURES

| | | |
|---|---|---|
| **FIGURE 2-1** | Definition of the operating window. | 15 |
| **FIGURE 4-1** | ITPM task selection process. | 31 |
| **FIGURE 5-1** | Training flow chart. | 62 |
| **FIGURE 6-1** | MI procedure hierarchy. | 81 |
| **FIGURE 6-2** | Basic procedure development process. | 85 |
| **FIGURE 8-1** | Technical evaluation condition selection. | 121 |
| **FIGURE 11-1** | Example risk matrix with ALARP region (Ref. 11-4). | 211 |
| **FIGURE 11-2** | Example RCM decision tree. | 217 |
| **FIGURE 11-3** | Management of risk using RBI. | 219 |
| **FIGURE 11-4** | RBI program flowchart. | 220 |
| **FIGURE 11-5** | Sample LOPA worksheet (Page 1 of 2). | 227 |
| **FIGURE 12-1** | MI program continuous improvement model. | 232 |
| **FIGURE 12-2** | Example MI process map with suggested performance measures. | 241 |
| **FIGURE 12-3** | Sample failure analysis process. | 244 |
| **FIGURE 12-4** | Sample fault tree. | 246 |
| **FIGURE 12-5** | Sample causal factor chart. | 246 |

# ITEMS ON THE CD ACCOMPANYING THIS BOOK

### Chapter 4 Resources

- *Imperfection vs. type of NDE method*, ASME Section V, Article 1, Nonmandatory Appendix A (reproduced with permission)
- Common RAGAGEPs
- Example condition monitoring (CM) techniques:
  - Temperature measurement
  - Dynamic monitoring
  - Oil analysis
  - Corrosion analysis
  - Nondestructive testing (NDT)
  - Electrical testing and monitoring
  - Observation and surveillance
  - Performance monitoring

### Chapter 5 Resources

- Sample training guide (repairing a mechanically sealed pump)
- Sample of skills/knowledge list for an electrician
- Sample of skills/knowledge list for a mechanic

### Chapter 6 Resources

- Sample procedures:
  - Mass Flowmeter Calibration
  - Level-Indicating Transmitter Function Test and Calibration
  - Heater Inspection
  - Contractor Safety
  - Circuit Breaker Inspection and Maintenance
  - Centrifugal Pump Specification Sheet

- Example procedure formats:
  - Flat procedure format
  - Outline procedure format
  - Playscript or information mapping procedure format
  - T-bar procedure format
  - Checklist procedure format
- Comparison of procedure formats
- Procedure writing checklist

**Chapter 9 Resources**

- Common RAGAGEPS
- Fixed equipment matrices:
  - Pressure vessel matrix, including columns, filters, and heat exchangers
  - Storage tanks
  - Process piping, and components, including buried piping, flex hoses, and expansion joints
- Relief and vent system matrices:
  - Pressure relief valves (PRVs)
  - Rupture discs
  - Vent headers
  - Flame/detonation arresters
  - Emergency vents
  - Thermal oxidizers
  - Flares
- Instrumentation and controls matrices:
  - Safety instrumented systems (SISs) and emergency shutdown (ESD) systems
  - Process control systems
  - Critical alarms and interlocks
  - Chemical monitors and detection systems
  - Conductivity, pH, and other process analyzers
  - Burner management systems
- Rotating equipment:
  - Pumps
  - Reciprocating compressors
  - Centrifugal compressors, including specific protection systems (e.g., pressure cutouts)
  - Process fans and blowers
  - Agitators and mixers
  - Electric motors
  - Gas turbines
  - Steam turbines
  - Gearboxes

ITEMS ON THE CD ACCOMPANYING THIS BOOK  xix

- Fired systems:
  - Heaters/furnaces/boilers
- Electrical systems:
  - Switchgear
  - Transformers
  - Motor controls
  - Uninterruptible power supplies (UPSs)
  - Emergency generators
  - Lightning protection
- Grounding systems

**Chapter 10 Resources**

- MI program development activity worksheet

**Chapter 11 Resources**

- Presentation papers related to analysis approaches:
  - *Risk-Based Approach to Mechanical Integrity Success on Implementation*
  - *An Insurer's View of Risk-Based Inspection*
  - *RCM Makes Sense for PSM-Covered Facilities*
  - *Lessons Learned from a Reliability-Centered Maintenance Analysis*

**Chapter 12 Resources**

- Resources for performing equipment failure analyses:
  - Additional detailed information on the analysis steps
  - An equipment failure analysis checklist
- Resources for performing root cause analyses:
  - SOURCE™ Investigator's Toolkit
  - Root cause map
  - Causal factor chart and fault tree templates
- MI program audit resources
- Presentation paper.—.*Improving Mechanical Integrity in Chemical and Hydrocarbon Processing Facilities.*—. *A Insurer's Viewpoint*

# ACRONYMS AND ABBREVIATIONS

| | |
|---|---|
| ACC | American Chemistry Council |
| ACCP | ASNT Central Certification Program |
| ACGIH | American Conference of Governmental Industrial Hygienists |
| AIChE | American Institute of Chemical Engineers |
| ALARP | as low as reasonably practicable |
| ANSI | American National Standards Institute |
| API | American Petroleum Institute |
| ASM | American Society for Metals (ASM International) |
| ASME | American Society of Mechanical Engineers |
| ASNT | American Society of Non-destructive Testing |
| ASTM | American Society of Testing and Materials (ASTM International) |
| AWS | American Welding Society |
| | |
| BPVC | Boiler and Pressure Vessel Code |
| | |
| CCPS | Center for Chemical Process Safety |
| CF | causal factor |
| *CFR* | *Code of Federal Regulations* |
| CI | Chlorine Institute |
| CM | condition monitoring |
| CMMS | computerized maintenance management system |
| CPI | chemical process industries |
| | |
| DOT | Department of Transportation |
| | |
| E&I | electrical and instrumentation |
| EPA | Environmental Protection Agency |
| ESD | emergency shutdown |

| | |
|---|---|
| FFS | fitness for service |
| FM | factory mutual research |
| FMEA | failure modes and effects analysis |
| FMECA | failure modes, effects, and criticality analysis |
| FTA | fault tree analysis |
| | |
| HAZMAT | hazardous material |
| HAZOP | hazard and operability |
| HI | Hydraulic Institute |
| | |
| IEC | International Electrotechnical Commission |
| IIAR | International Institute of Ammonia Refrigeration |
| IPL | independent protection layer |
| ISA | Instrumentation, Systems, and Automation Society |
| ISO | International Organization for Standardization |
| ITPM | inspection, testing, and preventive maintenance |
| | |
| LOPA | layer of protection analysis |
| | |
| MI | mechanical integrity |
| MOC | management of change |
| | |
| NB | National Board (of Boiler and Pressure Vessel Inspectors) |
| NBBPVI | National Board of Boiler and Pressure Vessel Inspectors |
| NBIC | National Board Inspection Code |
| NDE | nondestructive examination |
| NDT | nondestructive testing |
| NEC | National Electric Code |
| NFPA | National Fire Protection Association |
| | |
| OEM | original equipment manufacturer |
| OSHA | Occupational Safety and Health Administration |
| | |
| P&ID | piping and instrumentation diagram |
| PFD | probability of failure on demand |
| PHA | process hazard analysis |
| PM | preventive maintenance |
| PMI | positive material identification |
| PPE | personal protective equipment |

# ACRONYMS AND ABBREVIATIONS

| | |
|---|---|
| PRV | pressure relief valve |
| PSM | process safety management |
| PSSR | prestartup safety review |
| PSV | pressure safety valve |
| PWHT | postweld heat treatment |
| | |
| QA | quality assurance |
| QC | quality control |
| | |
| RAGAGEP | recognized and generally accepted good engineering practice |
| RBI | risk-based inspection |
| RCA | root cause analysis |
| RCM | reliability-centered maintenance |
| RMP | risk management program |
| ROI | return on investment |
| RP | recommended practice |
| | |
| SCBA | self-contained breathing apparatus |
| SCE | safety-critical equipment |
| SHE | safety, health, and environmental |
| SIF | safety instrumented function |
| SIL | safety integrity level |
| SIS | safety instrumented system |
| SME | subject matter expert |
| | |
| TML | thickness measurement location |
| | |
| UL | Underwriters Laboratories Inc. |
| UPS | uninterruptible power supply |
| USCG | United States Coast Guard |
| UT | ultrasonic thickness |

# GLOSSARY

*Acceptance criteria*   Technical basis used to determine whether equipment is deficient (e.g., when analyzing inspection, testing, and preventive maintenance [ITPM] results).

*Causal factor (CF)*   Equipment failure or human error that caused an incident or allowed incident consequences to be worse.

*Causal factor chart*   A sequence diagram that graphically depicts an incident from beginning to end; typically used to organize incident data and identify causal factors.

*Certification*   Completion of the formal training and qualification requirements specified by applicable codes and standards.

*Computerized maintenance management system (CMMS)*   Computer software for planning, scheduling, and documenting maintenance activities. A typical CMMS includes work order generation, work instructions, parts and labor expenditure tracking, parts inventories, and equipment histories.

*Condition monitoring (CM)*   Observing, measuring, and/or trending of indicators with respect to some independent parameter (usually time or cycles) to indicate the current and future ability of a structure, system, or component to function within acceptance criteria.

*Cost avoidance*   Return (often expressed in monetary terms) resulting from actions that prevent an incident from occurring.

*Critical operating parameter*   Process condition (e.g., flow rate, temperature) that can lead to an equipment failure if limits are exceeded.

*Damage/failure mechanism*   The mechanical, chemical, physical, or other process that results in equipment degradation. Identifying and inspecting for indications of the damage mechanism can be used to predict future failures.

*Decision tree*   A logic tree used in reliability-centered maintenance (RCM) to help determine the correct type of maintenance (e.g., predictive, preventive) to perform to reduce the likelihood of equipment failures.

*Equipment class*   A grouping of individual equipment items with similar design and operation, such that facilities should perform similar ITPM activities on all of the items.

*Equipment deficiency*   A condition that does not meet the acceptance criteria.

*Equipment failure analysis*   A systematic approach for analyzing equipment failures to determine the failure mechanism(s) and the root cause(s) that resulted in the failure.

*Failure mode*   A symptom, condition, or fashion in which equipment fails. A failure mode might be identified as loss of function, premature function (function without demand), an out-of-tolerance condition, or a simple physical condition such as a leak.

*Failure modes and effects analysis (FMEA)*   A systematic method for evaluating and documenting the causes and impacts of known types of equipment/component failures.

*Failure modes, effects, and criticality analysis (FMECA)*   A variation of FMEA that includes a quantitative estimate of the significance of the consequence of a failure mode.

*Fault tree*   A method for representing the logical combinations of failures that can lead to a specific main event (i.e., failure or accident of interest).

*Fitness for service (FFS)*   A systematic approach for evaluating the current condition of a piece of equipment in order to determine if the equipment item is capable of operating at defined operating conditions (e.g., temperature, pressure).

*Gantt chart*   A manner of depicting multiple, time-based project activities (usually on a bar chart with a horizontal time scale).

*Hazard and operability (HAZOP) analysis*   A systematic method in which process hazards and potential operating problems are identified using a series of guide words to investigate process deviations.

*Inspection, testing, and preventive maintenance (ITPM)*   Scheduled proactive maintenance activities intended to (1) assess the current condition and/or rate of degradation of equipment, (2) test the operation/functionality of equipment, and/or (3) prevent equipment failure by restoring equipment condition.

*ITPM program*   A management system that develops, maintains, monitors, and manages inspection, testing, and preventive maintenance activities.

*Layer of protection analysis (LOPA)*   A process of evaluating the effectiveness of independent protection layer(s) in reducing the likelihood or severity of an undesired event.

*Management of change (MOC)*   A management system for ensuring that changes to processes are properly analyzed (e.g., for potential adverse impacts), documented, and communicated to affected personnel.

*Mechanical integrity (MI)*   A management system for ensuring the ongoing durability and functionality of equipment.

*Nondestructive testing/examination (NDT/NDE)*   Evaluation of an equipment item with the intention of measuring an equipment parameter without damaging or destroying the equipment item.

*Operating window*   The parameters (i.e., safe upper and lower limits, run time) under which equipment can function without failure.

*Owner-user*   Person, plant, or corporation legally responsible for the safe operation of a pressure-retaining item (e.g., a pressure vessel).

GLOSSARY                                                                                                                   xxvii

***Performance measure*** A metric used to monitor or evaluate the operation of a program activity or management system.

***Positive material identification (PMI)*** The determination of the materials of construction of an equipment item or component (e.g., piping).

***Predictive maintenance*** An equipment maintenance strategy based on measuring the condition of equipment in order to assess whether it will fail during some future period, and then taking appropriate action to avoid the consequences of that failure.

***Prestartup safety review (PSSR)*** A management system for ensuring that new or modified processes are ready for startup by verifying that equipment is installed in a manner consistent with the design intent.

***Process hazard analysis (PHA)*** A systematic evaluation of process hazards with the purpose of ensuring that sufficient safeguards are in place to manage the inherent risks.

***Process safety information*** A compilation of chemical hazard, technology, and equipment documentation needed to manage process safety.

***Quality assurance (QA)*** Activities to ensure that equipment is designed appropriately and to ensure that the design intent is not compromised throughout the equipment's entire life cycle.

***Recommended and generally accepted good engineering practice (RAGAGEP)*** A document that provides guidance on engineering, operating, or maintenance activities based on an established code, standard, published technical report, or recommended practice (or a document of a similar name).

***Reliability-centered maintenance (RCM)*** A systematic analysis approach for evaluating equipment failure impacts on system performance and determining specific strategies for managing the identified equipment failures. The failure management strategies may include preventive maintenance, predictive maintenance, inspections, testing, and/or one-time changes (e.g., design improvements, operational changes).

***Remaining life*** An estimate, based on inspection results, of the time it will take for an equipment item to reach a defined retirement criterion (e.g., minimum wall thickness).

***Risk*** A measure of potential loss (e.g., human injury, environmental insult, economic penalty) in terms of the magnitude of the loss and the likelihood that the loss will occur.

***Risk analysis*** The development of a qualitative and/or quantitative estimate of risk based on engineering evaluation and mathematical techniques (quantitative only) for combining estimates of event consequences and frequencies.

***Risk-based inspection (RBI)*** A systematic approach for identifying credible failure mechanisms and using equipment failure consequences and likelihood to determine inspection strategies for equipment.

***Root cause analysis (RCA)***  A method used to (1) describe what happened leading up to and during a particular event, (2) determine how it happened, and (3) identify the underlying reasons the event was allowed to occur so that workable corrective actions can be implemented to help prevent recurrence of the event (or occurrence of similar events).

***Safety instrumented functions (SIF)***  A system composed of servers, logic servers, and final control elements for the purpose of taking the process to a safe state when predetermined conditions are violated.

***Safety instrumented system (SIS)***  One or more SIFs combined to protect a process or a key process component.

***Safety integrity level (SIL)***  Criterion defining the acceptable probability of failure on demand for a safety instrumented system.

***Technical assurance***  The process for communicating that appropriate technology is being applied to process equipment.

***Technical evaluation condition***  An equipment condition requiring further technical evaluation to determine suitability for continued service.

***Verification activity***  A test, field observation, or other activity used to ensure that personnel have acquired necessary skills and/or knowledge following training.

# ACKNOWLEDGMENTS

The Chemical Center for Process Safety (CCPS) thanks all of the members of the Mechanical Integrity (MI) Subcommittee for providing technical guidance in the preparation of this book. CCPS also expresses its appreciation to the members of the Technical Steering Committee for their advice and support.

The chairmen of the MI Subcommittee were Brian Dunbobbin of Air Products and Chemicals and Thomas Folk of Rohm and Haas Company. The CCPS staff liaison was Dan Sliva. The Subcommittee had the following additional members:

John Alderman
RRS Engineering

Jon Batey
Dow Chemical Company

Gavin Floyd
Rhodia

Stan Grabill
Honeywell

Michael Hazzan
AcuTech Consulting Group

Dan Long
Celanese

Michael Moriarty
Akzo Nobel

Henry Ozog
IoMosaic

Chris Payton
BP

ABSG Consulting Inc. (ABS Consulting) prepared these MI guidelines. Randal Montgomery and Andrew Remson were the lead authors. Steve Arendt, William Bradshaw, Earl Brown, Myron Casada, Douglas Hobbs, and Daniel Machiela also contributed to these guidelines.

The authors of *Guidelines for Mechanical Integrity Systems* are indebted to the technical publications personnel at ABS Consulting. Karen Taylor was the editor for the manuscript. Paul Olsen created many of the graphics. Finally, Erica Suurmeyer prepared the copy for peer review and Jennifer Trudeau and Susan Hagemeyer prepared the final manuscript for publication.

CCPS also gratefully acknowledges the comments submitted by the following peer reviewers:

| | |
|---|---|
| Michael Altmann<br>Sunoco Inc. | Lisa Morrison<br>PPG |
| Mike Broadribb<br>BP | Jim Muoio<br>Lyondell Petrochemicals |
| David Cummings<br>Du Pont | Jack Philley<br>Baker Risk |
| Art Dowell<br>Rohm and Haas Company | Bill Salot<br>Honeywell |
| Cindy Gross<br>Celanese | Mark Saunders<br>Georgia Pacific |
| Peter Howell<br>Mark V, Inc. | Kenan Stevick<br>Dow Chemical Company |
| Pete Lodal<br>Eastman Chemical Company | Jim Willis<br>Air Products and Chemicals |
| Bill Marshall<br>Eli Lilly | |

Their insight and suggestions helped ensure a balanced perspective for the *book*.

# PREFACE

The American Institute of Chemical Engineers (AIChE) has been closely involved with process safety and loss control issues in the chemical and allied industries for more than four decades. Through its strong ties with process designers, constructors, operators, safety professionals, and members of academia, AIChE has enhanced communications and fostered continuous improvement of the industry's high safety standards. AIChE publications and symposia have become information resources for those devoted to process safety and environmental protection.

AIChE created the Center for Chemical Process Safety (CCPS) in 1985 after the chemical disasters in Mexico City, Mexico, and Bhopal, India. The CCPS is chartered to develop and disseminate technical information for use in the prevention of major chemical accidents. The center is supported by more than 80 chemical process industries (CPI) sponsors who provide the necessary funding and professional guidance to its technical committees. The major product of CCPS activities has been a series of guidelines to assist those implementing various elements of a process safety and risk management system. This book is part of that series.

Mechanical integrity (MI) is a fundamental component of successful process safety programs. However, facilities continue to be challenged to maintain successful MI programs. The CCPS Technical Steering Committee initiated the creation of these guidelines to assist facilities in meeting this challenge. This book contains approaches for designing, developing, implementing, and continually improving an MI program. The CD accompanying this book contains resource materials and support information.

# MANAGEMENT OVERVIEW OF THE GUIDELINES

Mechanical integrity (MI) is a product of many activities, usually performed by many people. When these activities are done well, MI can provide the foundation for a safe, reliable facility that minimizes threats to the environment, the public, and the workforce. These factors, as well as regulatory drivers, provide ample motivation for ensuring appropriate MI. Simply put, quality MI is consistent with good business practices.

Quality MI is an integrated product of proper equipment, dependable human performance, and effective management systems. *Guidelines for Mechanical Integrity Systems (MI Guidelines)* walks the reader through the development, implementation, and continual improvement of an MI program that includes these areas of focus. Behind these focus areas must be an involved, supportive management. These guidelines also include advice to those supporting the program.

To begin developing an MI program, facility management needs to decide which equipment items are to be included and at what level of detail. Similarly, improving an existing MI program requires confirmation that these decisions were made correctly and are implemented consistently. Thus, these guidelines are written for a wide range of audiences and potential users. Furthermore, although compliance with federal, state, and local regulations is often a motivating factor for a facility, the *MI Guidelines* are not limited by any regulations. Rather, following the *MI Guidelines* should help a facility develop, implement, and/or improve an effective MI program and also be in compliance with those regulations.

Once the goals, objectives, and equipment definitions are determined, facility management needs to develop and implement systematic activities related to MI. These include:

- Inspection, testing, and preventive maintenance (ITPM)
- Training of all affected personnel
- Relevant procedures
- Quality assurance (QA)
- Equipment deficiency resolution

The *MI Guidelines* contains approaches for all of these. Specific details for these activities should depend on facility culture, regulatory obligations, and company priorities. Therefore, relatively little prescriptive information is included in this book. Rather, the *MI Guidelines* present approaches that have worked in facilities of various industries and various sizes.

To a large extent, the implementation of MI activities depends on the specific types of equipment involved. Because many codes, standards, and other guidelines are written for specific equipment types, the *MI Guidelines* contains a section dedicated to the specific approaches applicable to different equipment types. Many recognized and generally accepted good engineering practices (RAGAGEPs) available are listed in this section. Key aspects of the RAGAGEPs (e.g., time intervals between inspections) are also listed, but the reader is encouraged to consult the referenced documents for more detailed information.

Implementation of an effective MI program usually requires extensive resources. Frequently, these resources include a computerized maintenance management system (CMMS). Many commercial CMMS packages are available, although some in-house programs are also effective. The *MI Guidelines* includes basic information that should be included in any CMMS that is installed as part of an effective MI program.

Because of the extensive resource requirements for most MI programs, risk-based decision making can be effectively employed to prioritize resource allocation. Various texts have been written on applicable tools for making these decisions. The *MI Guidelines* includes an overview of many of the tools and references available in these resources.

The *MI Guidelines* closes with a section on continuous improvement of MI systems. Continuous improvement is needed to ensure that effective MI programs continue to operate at a high level. Some improvement can be attained simply by asking the right questions (i.e., auditing) and following up to address any problems. Performing improvement activities on a frequent basis will lead to continuous improvement.

For the convenience of the reader, appendix materials and bibliographical references are listed at the conclusion of each section of this book. In addition, materials that are (1) too voluminous to efficiently include in an appendix; (2) directly copied, with permission, from other sources; and/or (3) desirable to use in an electronic format, are included in the CD that accompanies this book.

Finally, the reader and management should be aware that effective MI programs cannot guarantee that incidents will not occur. However, an effective MI program, in conjunction with other effective risk management programs (RMPs), can reduce the risk associated with chemical operations.

# 1
# INTRODUCTION

For decades, mechanical integrity (MI) activities have been a part of industry's efforts to prevent incidents and maintain productivity. Industry initiatives, company initiatives, and regulations in various countries have helped (1) define MI program requirements and (2) accelerate implementation of MI programs. MI is already ingrained in the culture of many process plants, as well as in other related industries. Some MI activities are essential for these facilities to maintain economic viability.

Since 1992, a major incentive for the chemical process industries (CPI) in the United States to implement MI programs has been the Occupational Safety and Health Administration's (OSHA's) process safety management (PSM) regulation (29 *Code of Federal Regulations* [CFR] 1910.119) (Reference 1-1). This was followed by the Environmental Protection Agency's (EPA's) risk management program (RMP) rule (40 CFR 68) (Reference 1-2). These performance-based regulations each contain an MI element that defines the minimum requirements of a program through six subelements that address:

- Application (equipment to include)
- Written procedures
- Training
- Inspection and testing
- Equipment deficiencies
- Quality assurance (QA)

The specific requirements are not prescriptively stated in these regulations, but the subelements represent time-proven practices for an effective MI program. The details of each subelement are left to the discretion of the facility to develop and implement. All PSM- and RMP-covered U.S. facilities in operation since the regulations were issued should have completed at least three compliance audits. Many of these audits reveal that companies continue to have significant opportunities to improve their MI programs. In response, the Center for Chemical Process Safety (CCPS) Technical Steering Committee launched a project to develop a guidance book to address the development, implementation, management, and continuous improvement of MI programs.

This guidelines book was written for CPI companies; however, the majority of the information presented applies to other industries as well. Although this book was written in the United States, a conscious effort has been made to keep the book applicable to facilities worldwide. This book recommends efficient approaches for establishing a successful MI program, while taking into consideration that facilities with small staffs and fewer resources must also develop MI programs. The practices described in this book are intended to help facilities create or improve MI programs.

## 1.1 WHAT IS MECHANICAL INTEGRITY?

For the purposes of this book, MI is the programmatic implementation of activities necessary to ensure that important equipment will be suitable for its intended application throughout the life of an operation. MI programs vary according to industry, regulatory requirements, geography, and plant culture. However, some characteristics are common to all good MI programs. For example, a successful MI program:

- Includes activities to ensure that equipment is designed, fabricated, procured, installed, operated, and maintained in a manner appropriate for its intended application
- Clearly designates equipment included in the program based on defined criteria
- Prioritizes equipment to help optimally allocate resources (e.g., personnel, money, storage space)
- Helps a plant staff perform planned maintenance and reduce the need for unplanned maintenance
- Helps a plant staff recognize when equipment deficiencies occur and includes controls to help ensure that equipment deficiencies do not lead to serious accidents
- Incorporates recognized and generally accepted good engineering practices (RAGAGEPs)
- Helps ensure that personnel assigned to inspect, test, maintain, procure, fabricate, install, decommission, and recommission process equipment are appropriately trained and have access to appropriate procedures for these activities
- Maintains service documentation and other records to enable consistent performance of MI activities and to provide accurate equipment information to other users, including other process safety and risk management elements

This book provides advice for developing an MI program with all of these characteristics.

# 1. INTRODUCTION

## 1.2 RELATIONSHIP TO OTHER PROGRAMS

A practical MI program will fit within a facility's existing process safety and RMPs as well as other improvement initiatives (e.g., reliability, quality). Personnel charged with developing and administering the MI program can optimize the process by taking advantage of existing programs and by knowing which people and groups of people are responsible for related activities. Table 1-1 illustrates potential interfaces with other facility programs.

## 1.3 EXPECTATIONS FOR THE MI PROGRAM

To present sound guidance for developing and/or improving MI programs, this guidelines book evaluated lessons learned by the CPI. This is not a "cookbook"; however, many ways to approach the implementation of an MI program exist. MI programs must be effective in preventing incidents and should be an efficient component of a facility's process safety, environmental, risk, and reliability management system(s). Where appropriate, this book presents strengths and weaknesses of different approaches. Company management will need to recognize which approaches best suit their facility and company needs.

One beneficial practice is to establish program objectives early in the MI program development process. Companies should consider the implications of setting objectives for their programs. Reasonable expectations of MI programs include:

- Improved equipment reliability
- Reduction in equipment failures that lead to safety and environmental incidents
- Improved product consistency
- Improved maintenance consistency and efficiency
- Reduction of unplanned maintenance time and costs
- Reduced operating costs
- Improved spare parts management
- Improved contractor performance
- Compliance with government regulations

However, each of these objectives may have associated costs (e.g., more detailed procedures, a larger warehouse, improved computer systems); therefore, companies should consider prioritizing their objectives.

One MI program development approach that is not advocated in this book is to focus on compliance with regulations. The motivation for this approach is usually financial. Unfortunately, using this philosophy often puts a facility at a disadvantage because the requirements for compliance are often vague and subject to misinterpretation. Furthermore, requirements are subject to change (via legislated modifications or new interpretations of existing legislation). In addition, a compliance-only program may miss out on many of the benefits of a more

## TABLE 1-1
## Potential MI Interfaces with Other Programs

| Program | Potential MI Interface |
|---|---|
| Equipment Reliability | • Reliability program activities (e.g., vibration monitoring, equipment quality control [QC]) contribute to MI<br>• An MI program can be the foundation of a plant's reliability program |
| Occupational Safety | • Occupational safety programs help ensure the safe performance of MI activities<br>• Occupational safety personnel may help maintain the integrity of emergency response equipment |
| Environmental Control | • Environmental initiatives (e.g., monitoring for fugitive emissions, investigating chemical releases) contribute to MI |
| Employee Participation | • Employees from various departments should have input into the MI program |
| Process Safety Information | • Design codes and standards influence MI activities such as equipment design, inspection, and repair<br>• MI QA activities help document that equipment is appropriate for its intended use<br>• MI activities may help establish or dictate a change to safe upper and lower operating limits |
| Process Hazard Analysis (PHA) | • PHAs can help define the equipment scope for the MI program<br>• PHAs can help prioritize MI activities<br>• MI history can help PHA teams determine the adequacy of safeguards |
| Operating Procedures | • Operating procedures may cover MI-related activities, such as equipment surveillance as part of operator rounds, reporting operating anomalies, recording historical equipment operating data, and preparing equipment for maintenance |
| Operator Training | • MI training in an overview of the process and its hazards should be consistent with the content of the operator training program |
| Contractors | • Inspection and maintenance tasks under the MI program may dictate skills required of contractors<br>• Because contractors often perform MI activities, the contractor selection process should consider both contractor safety performance and the quality of the contractor's work |
| Prestartup Safety Review (PSSR) | • The MI QA practice to ensure that equipment is fabricated and installed according to design may be fully or partially addressed during a PSSR |
| Hot Work Permit (and other safe work practices) | • Safe work practices are relied upon to perform MI activities |
| Management of Change (MOC) | • MOC should apply to MI activities and documents (e.g., changes to task frequencies, procedures)<br>• The MOC program should ensure that MI issues (e.g., corrosion rates and mechanisms) are considered when evaluating process changes<br>• Establish hazard review teams that include process and MI personnel<br>• The MOC program may be upgraded to help manage equipment deficiencies<br>• Practices for replacing equipment "in kind" should be reviewed to ensure that MI records are not compromised (e.g., inspection records and schedules are updated) |
| Incident Investigation | • MI records may be needed by investigation teams<br>• Investigation recommendations may impact MI activities |
| Emergency Planning and Response | • Emergency response equipment should be included in the MI program |
| Compliance Audits | • The MI program will be audited — audit results can help improve the MI program |
| Trade Secrets | • Trade secrets needed for MI activities cannot be withheld |

# 1. INTRODUCTION

holistic approach, such as reduced risks for employees, the neighboring community, and the facility. A more holistic approach can help to:

- Present the MI program as a company priority, rather than just something the company is forced to do; this approach helps to ensure compliance because personnel are less likely to take shortcuts
- Create synergies with equipment and process reliability initiatives that could improve results and/or lower cost
- Address actual risks to employees, community, and the business

Therefore, the more holistic approach helps to ensure compliance with governing regulations and, ultimately, often turns out to be less expensive than the minimum compliance effort would have been.

## 1.4 THE EFFECT OF RAGAGEPs

RAGAGEPs are important resources for an MI program. Many process safety reference documents and guidance documents rely on RAGAGEPs for a wide range of equipment and practices. For example:

- CCPS, "Design codes represent ... minimum requirements"; *Guidelines for Engineering Design for Process Safety* (Reference 1-3).
- CCPS, "The more widely accepted design practices are contained in various national and industry standards"; *Guidelines for Implementing Process Safety Management Systems* (Reference 1-4).
- American Chemistry Council (ACC), "Each member company shall have an ongoing process safety program that includes ... facility design, construction, and maintenance using sound engineering practices consistent with recognized codes and standards"; *Resource Guide for the Process Safety Code of Management Practices for Facilities*, Responsible Care® Process Safety Code (Reference 1-5).

In addition, regulations require the use of RAGAGEPs:

- EPA and OSHA, "Inspection and testing procedures shall follow recognized and generally accepted good engineering practices"; EPA 40 CFR 68 and OSHA 29 CFR 1910.119.
- EPA and OSHA, "The employer (owner or operator) shall document that equipment complies with recognized and generally accepted good engineering practices"; EPA 40 CFR 68 and OSHA 29 CFR 1910.119.

What are RAGAGEPs? Simply stated, RAGAGEPs are documents that provide guidance on engineering, operating, or maintenance activities based on an established code, standard, published technical report, or recommended practice (RP) (or a document of a similar name) (Reference 1-6). They outline in detail a generally approved way to perform a specific engineering, inspection, or maintenance activity, such as fabricating a pressure vessel, inspecting a storage tank, or servicing a relief valve. Many of these documents were developed after obtaining broad industry and expert public technical input, and many were accepted by consensus of industry and technical organizations. Therefore, RAGAGEPs provide a valuable starting point for an MI program.

In some cases, a country, state, or locality may mandate the use of a RAGAGEP. In addition, many companies internalize standards provided by the manufacturer or licensor of a process (these often are based on RAGAGEPs). Some companies have developed internal standards based on company and industry operating experience. To effectively use RAGAGEPs, facility management must determine which practices are available and then assess the applicability of each practice to their facility. Regardless of the consensus reached to publish a RAGAGEP, most standards were not written for a facility's specific equipment, specific chemical application, specific locale, or specific operations culture. Facilities with successful MI programs are establishing their own data records to help determine (or to validate) the ongoing applicability and use of each standard.

Several chapters of this book address the applicability and use of RAGAGEPs in more detail. Descriptions of these practices, and approaches for using them (e.g., to determine an inspection interval or technique), are included, but the actual RAGAGEPs are not repeated in this book. New and revised codes, standards, and recommended practices continue to evolve; therefore, companies should have management systems in place to keep up with the new standards and with changes to existing standards.

## 1.5 STRUCTURE OF THIS GUIDELINES BOOK

This guidelines book is intended for anyone interested in developing a new MI program or enhancing an existing program. The book was written in the United States, but few references are made to jurisdictional regulations. Note, however, that following the approaches described in the book should help any organization that is trying to comply with rules or regulatory requirements related to MI. Similarly, the codes and standards referenced are generally from the United States, but other code references are also provided. Also note that most of the information within the book is consistent with codes and regulations, but it is not extracted directly from those codes or regulations. The suggested approaches are applicable everywhere.

This guidelines book begins with chapters that help set the groundwork for the MI program. Chapter 2 discusses roles and responsibilities for company personnel and examines the ongoing activities that management undertakes to help ensure MI program success. Chapter 3 reviews considerations a facility may have when defining the equipment to include in its program.

Chapter 4 discusses inspection, testing, and preventive maintenance (ITPM). Some peer reviewers of this text suggested that preventive maintenance (PM) does not belong in an MI program. Many traditional PM programs were established to

# 1. INTRODUCTION

address routine nonintegrity-related tasks. However, in this book "preventive maintenance" refers to those activities performed to prevent the failure of equipment within the MI program that are not inspections or tests (e.g., lubrication of rotating equipment).

Chapter 5 covers personnel training and Chapter 6 addresses the procedures needed for MI. A life-cycle approach to QA is presented in Chapter 7. Chapter 8 covers equipment deficiency recognition and resolution. Chapter 9 is dedicated to the equipment-specific aspects for the management systems covered in Chapters 4 through 8. Chapter 10 reviews common issues encountered with MI program implementation. The remaining two chapters contain supplemental information related to MI programs. Chapter 11 provides overviews of risk-based tools that can be used to help make decisions related to MI activities. Chapter 12 offers advice for continual assessment and improvement of an MI program. Many MI activities are concentrated in four areas:

1. New equipment (design, fabrication, and installation)
2. Inspection and testing
3. Preventive maintenance
4. Repair

As illustrated in Table 1-2, Chapters 4 through 8 describe management systems for addressing these four areas. Chapter 9 is dedicated to the equipment-specific aspects for these areas. Activity tables in Chapter 9 and on the CD accompanying this book are presented in a format similar to Table 1-2.

**TABLE 1-2**
**Chapters Addressing Management Systems for MI Activities**

| Attributes | New Equipment | Inspection and Testing | Preventive Maintenance | Repair |
|---|---|---|---|---|
| Task Definition, Purpose, and Documentation Requirements | Chapter 7 (QA) | Chapter 4 (ITPM) | Chapter 4 (ITPM) | Chapter 8 (Deficiency Resolution) |
| Acceptance Criteria | Chapter 7 (QA) | Chapter 4 (ITPM) and Chapter 8 (Deficiency Resolution) | Not applicable | Chapter 7 (QA) |
| Technical Basis | Chapter 7 (QA) | Chapter 4 (ITPM) | Chapter 4 (ITPM) | Chapter 7 (QA) |
| Procedures | Chapter 6 (MI Procedures) | | | |
| Personnel Qualifications | Chapter 5 (MI Training) | | | |

## 1.6 REFERENCES

1-1 Occupational Safety and Health Administration, *Process Safety Management of Highly Hazardous Chemicals*, 29 CFR Part 1910, Section 119, Washington, DC, 1992.

1-2 Environmental Protection Agency, *Accidental Release Prevention Requirements: Risk Management Programs*, Clean Air Act, Section 112 (r)(7), Washington, DC, 1996.

1-3 American Institute of Chemical Engineers, *Guidelines for Engineering Design for Process Safety*, Center for Chemical Process Safety, New York, NY, 1993.

1-4 American Institute of Chemical Engineers, *Guidelines for Implementing Process Safety Management Systems*, Center for Chemical Process Safety, New York, NY, 1994.

1-5 American Chemistry Council, *Resource Guide for the Process Safety Code of Management Practices*, Washington, DC, 1990.

1-6 Decker, L. and R. Montgomery, *Defining and Maintaining a RAGAGEP Program*, presented at Process Plant Safety Symposium, Houston, TX, December 1998.

# 2
# MANAGEMENT RESPONSIBILITY

Many people within a facility's maintenance, operations, and engineering organizations should be involved with the facility's mechanical integrity (MI) program. An individual's involvement may range from brief encounters to career-length stewardship responsibility, and the involvement may occur during any or all phases of the equipment's life cycle. In successful MI programs, supervisors and managers emphasize how each person contributes to preventing incidents and to improving process reliability. Such an approach is evident when personnel are working within the facility's hazard management system and using effective knowledge, skills, resources, and procedures associated with the MI program.

This chapter discusses ways in which supervisors and managers can contribute to the ultimate success of the MI program through communication and effective application of knowledge, skills, and resources. For the purposes of this chapter, supervisors and managers include all personnel with supervisory and/or management responsibilities within engineering, maintenance, operations, and related departments, as well as the personnel charged with overall facility leadership.

## 2.1 FACILITY LEADERSHIP'S ROLES AND RESPONSIBILITIES

One of the best ways that a facility's leadership can help prevent incidents is to provide visible and active involvement in the facility's hazard management system. Key responsibilities of managers and supervisors in the MI program are to (1) ensure that knowledgeable people are performing appropriate activities using effective engineering and decision-making tools and methods, (2) instill the expectation that the business plan will be fulfilled only within the safe operating limits of the equipment as dictated by its condition, (3) ensure that MI program activities (e.g., inspections and tests) are being executed and managed on schedule and as planned, and (4) ensure that appropriate controls are implemented and maintained within the facility's hazard management system for all related MI activities. The primary control mechanisms for these actions are:

- Establishing clear organizational roles, responsibilities, and accountability for MI activities
- Creating reporting mechanisms for equipment condition and for MI program status
- Ensuring that effective audits of the MI program and the overall hazard management system are conducted

### 2.1.1 Organizational Roles and Responsibilities

An essential role for a facility's management is to ensure that knowledgeable people are available and assigned to provide the expertise for implementing the MI program. For example, technical personnel at the facility or available to the facility should be able to direct personnel from various departments to the correct recognized and generally accepted good engineering practice(s) [RAGAGEP(s)] for a given application throughout the life cycle of the facility's equipment. The life cycle includes the process equipment's design/engineering, fabrication, procurement, receipt, storage and retrieval, construction and installation, commissioning, operation, inspection, condition monitoring (CM), function testing, maintenance, repair, modification and alteration, decommissioning, removal, and reuse.

Management personnel can demonstrate further commitment to the program by supporting the recommendations made by the technical personnel and by not inappropriately overriding those recommendations with operational or economic constraints. Facility leadership can provide the guidance and direction to ensure that the equipment technical roles are accepted by everyone within the facility. For example, as custodians of the equipment's condition-based safe operating limits, the technical staff should have significant input into decisions regarding whether equipment with known deficiencies should continue to be operated. The technical staff provides the input that enables facility management to manage the facility assets while maintaining safe operations.

### 2.1.2 Roles and Responsibilities Matrix

The roles and responsibilities for program management and implementation can be assigned to personnel in various departments. A thorough program document effectively communicates the management systems and roles associated with the MI program to facility personnel. In fact, many facilities find it a practical necessity to create and maintain a more detailed written MI procedure in order to sustain all MI activities. The written program can document roles and responsibilities for various levels of involvement for different aspects of the MI program. Some people will be directly responsible and accountable for an activity, others may participate in the development or implementation of the activity, and others may need only to be aware of the activity or the results from that activity. A convenient method for illustrating these varying roles and responsibilities is a roles/responsibility matrix. Such a matrix correlates different activities with different job positions by indicating the job position(s) responsible for each activity, as well as the level of participation required of other personnel.

## 2. MANAGEMENT RESPONSIBILITY

Example roles and responsibilities for MI program management are provided in Table 2-1. This matrix lists program activities in the left column and typical job positions by department in the top row. Cells containing letter designators indicate the job position(s) responsible for the activity and the levels of participation for other job positions. This matrix uses the following letter designators:

- R indicates the job position(s) with primary responsibility for the activity.
- A indicates the job position(s) to approve the work or decisions involved.
- S indicates the job positions that typically support the responsible person(s) and may participate in the performance of the activity.
- I indicates the job positions that are likely to be informed of the activity results, may be asked to provide information, or may have minor participation in the activity.

This matrix format will be used to present more specific roles and responsibilities in subsequent chapters of this book.

### 2.1.3 Reporting Mechanisms

Reporting mechanisms are necessary to provide appropriate and timely information concerning past performance as well as to alert personnel at various levels within the company that MI activities are necessary to ensure continued equipment integrity. Examples of information of interest to personnel working in the facility would include answers to such questions as:

- How sound is the equipment?
- Will the equipment continue to be fit for service tomorrow, next week, and next year?
- Are inspections overdue? How much scheduled maintenance is on the backlog? Are there unresolved equipment deficiencies?
- What information are the measures of equipment condition providing?
- Are MI-related incidents and near misses consistently reported?
- Are qualified people available to perform all required MI tasks?
- Are engineered safeguards always effective?
- Are MI procedures and standards consistently followed?
- What information are MI metrics providing?
- With what confidence can MI data be used?

**TABLE 2-1**
**Roles and Responsibilities Matrix for MI Program Management**

| Activity | Facility Leadership/ Site Manager | Engineering Manager/ Lead (Technical Assurance Role) | Maintenance Manager | Maintenance Supervisors | Area Superintendent/ Unit Manager | Production Supervisors | Process Engineering Manager | EHS Manager | PSM Coordinator |
|---|---|---|---|---|---|---|---|---|---|
| Program Coordination and Overall Responsibility | A | S | R | S | S | — | — | — | — |
| Establish Audit Schedule and Content | A | S | S | — | — | — | — | R | R |
| Develop Metrics | — | R | A | S | S | S | S | — | — |
| Report/Review Metrics | A | R | R | — | S | — | S | — | — |
| Establish Organizational Roles | R | S | S | — | S | — | S | — | — |
| Provide Technical Content |   | R | S | — | S | — | S | — | — |
| Training/Personnel Development | A | R | R | S | S | — | S | S | — |

*(Note: Activities for which responsibility is assigned to more than one job position indicate shared responsibility or responsibility for a specific scope within the activity.)*

## 2. MANAGEMENT RESPONSIBILITY

### TABLE 2-1 (Continued)

| Activity | Facility Leadership/ Site Manager | Engineering Manager/Lead (Technical Assurance Role) | Maintenance Manager | Maintenance Supervisors | Area Superintendent/ Unit Manager | Production Supervisors | Process Engineering Manager | EHS Manager | PSM Coordinator |
|---|---|---|---|---|---|---|---|---|---|
| Overseeing Contractors in MI Roles | A | P | R | P | P | — | — | — | — |
| Equipment Deficiency Management | — | R | S | — | R | S | S | S | — |
| Procedure Development | — | R | A | S | — | — | — | — | — |
| Define/Maintain RAGAGEPs | — | R | A | — | — | — | — | — | — |
| Establish Safe Design Limits | — | A | — | — | S | S | R | — | — |
| Maintain Equipment Records | — | R | A | S | — | — | — | — | — |
| Establish Quality Assurance (QA) Requirements | — | R | A | — | — | — | — | — | — |

*(Note: Activities for which responsibility is assigned to more than one job position indicate shared responsibility or responsibility for a specific scope within the activity.)*

Management at facilities with effective MI programs ensures that (1) these and similar questions are being asked periodically and (2) the answers are complete and accurate. Although a facility may have experienced years of incident-free operations, complacency toward procedure adherence and less-than-thorough auditing of the hazard management system, which may have allowed the development of poor maintenance and operating practices, can result in undesirable answers to the questions above.

Management should be routinely informed of the answers to questions such as these to help ensure equipment integrity. Because every operation and its associated hazards are unique, facility management should work with operations personnel to create appropriate questions for each process.

### 2.1.4 Auditing

Auditing is an important aspect of any management system. Formal and informal auditing provides a means for facility management to understand the effectiveness of the MI activities. Some level of MI auditing activities should occur on a continual basis within the facility. These audit activities may include routine walk-about discussions, periodic stewardship/status reports, regulatory-required audits, periodic technical audits, and departmental effectiveness audits. Specific details on MI program auditing are provided in Chapter 12.

Auditing is valuable because the continuous nature of MI activities generates much information that forms the basis of day-to-day decisions. Understanding the equipment integrity issues can lead to successful and incident-free operation. Ongoing and continual auditing practices can help uncover equipment integrity problems or MI program deficiencies before an incident occurs.

## 2.2 TECHNICAL ASSURANCE RESPONSIBILITIES

Technical management and supervisory personnel ensure that the MI program activities are completed in accordance with the requirements of the hazard management system and in a manner that meets the requirements of the specific tasks. Therefore, management must ensure that qualified personnel are available to perform all tasks. In some cases, qualified personnel can be drawn from central company organizations. However, having staff members on site who are qualified to perform the necessary MI tasks is often preferable. (Chapter 5 discusses training and qualifying personnel to perform MI tasks.) The following sections discuss some specific technical assurance responsibilities to be provided in support of the MI program.

### 2.2.1 Defining Acceptance Criteria

Management personnel are responsible for defining appropriate acceptance criteria for equipment. These criteria include the acceptable operating window over the life of the equipment in practical terms of operating limits (process variable or measured material limits) and run time before the next condition inspection, function test, repair, or replacement. The upper and lower limits of the

## 2. MANAGEMENT RESPONSIBILITY

operating window are based on both the equipment's current condition or functionality assessment and its projected condition or functionality over time. Figure 2-1 illustrates this operating window.

Each MI activity must satisfy acceptance criteria and be consistent with RAGAGEPs. Supervisory and technical management personnel facilitate these activities by employing technical assurance personnel to (1) develop procedures and other MI documents with technical content requirements, (2) determine metrics for equipment CM and MI program performance, and (3) technically review the results of MI activities.

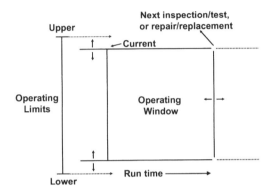

FIGURE 2-1  Definition of the operating window.

### 2.2.2  Providing Technical Content

The technical content requirements to be provided should include (1) details of how work is to be performed or checked, (2) requirements for materials of construction, (3) applicable codes and standards to be followed, (4) acceptance criteria for inspections and tests, and (5) equipment-specific inspection techniques. In addition, decisions to operate equipment with a deficiency can be based on data from technical assurance personnel. Such information may include the equipment's current and predicted condition, repair requirements, on-stream monitoring requirements, and/or revised safe operating limits. See Chapter 8 for information on managing deficient equipment.

### 2.2.3  Establishing Metrics

Technical assurance personnel can identify relevant standards of measurement, or metrics, as indicators of equipment integrity. A facility's metrics should include measures related to (1) MI program implementation and schedule adherence, (2) equipment condition trends, (3) adherence to procedures, and (4) training status. Discipline-specific measures can provide the information needed to communicate the status of equipment integrity activities and trends that affect equipment integrity (e.g., increase in past-due inspections and tests). Typically, metrics are used to report the status of such MI concerns as:

- MI activities backlog management
- Preventive maintenance (PM) schedule adherence
- Equipment deficiency resolution
- Each equipment type's inspection and testing schedule adherence
- Recommendation resolution and implementation (e.g., from inspection and testing activities)
- Equipment incident investigation recommendation resolution and implementation
- Equipment operating near design limits or life
- MI program audit findings and recommendation resolution and remediation
- MI procedure use (e.g., are the procedures current?)
- Craft and technician training and qualification

Using metrics is further discussed in Chapter 12.

### 2.2.4 Ensuring Technical Review

Technical review of MI activity results is a critical technical assurance role. The volume of data gathered and the potential for rapidly changing equipment conditions necessitate timely review of the inspection, testing, and preventive maintenance (ITPM) and equipment repair results. Prompt technical review of such results (1) identifies equipment with deficient or near-deficient conditions, (2) allows rapid communication to affected personnel regarding the unavailability of safeguards, (3) improves timing and effectiveness of corrective actions. and (4) helps prevent incorporating incorrect information into equipment history records.

# 3
# EQUIPMENT SELECTION

Early in the development of a mechanical integrity (MI) program, facility personnel need to establish boundaries and develop a list of the equipment to include. This chapter discusses some criteria to consider when identifying the equipment to include in the MI program. Addressing these items will help to ensure that (1) the program includes all desired equipment and (2) the basis for the program is consistently understood. Some of the steps that facilities take to establish equipment scope include:

- Review the program objectives.
- Establish and document the equipment selection criteria (i.e., which types of equipment to include or exclude).
- Define the level of detail (e.g., whether equipment is to be included individually or only as part of a system).
- Document the equipment selected.

This effort produces an equipment list that can serve as (1) the basis for establishing inspection and testing plans, (2) record – filing systems, and (3) other equipment-specific components of the MI program.

## 3.1 REVIEWING PROGRAM OBJECTIVES

Facility personnel should review the program objectives before establishing the equipment scope. Is the facility covered by process safety management (PSM), the risk management program (RMP) rule, state boiler and/or pressure vessel rules, or any other regulations? If so, the jurisdiction may dictate specific equipment that is required to be included in the MI program. Regulations may also be used to limit the extent of program coverage. However, be aware that regulations may be vague and interpretations of the regulations may change; therefore, MI programs seeking only compliance may be less effective than those with additional objectives (see Section 1.3).

Company management is motivated to develop MI programs for reasons that may extend beyond regulatory compliance. Reducing the likelihood of process safety incidents and/or occupational safety or environmental incidents are common incentives for including additional equipment in the program. Product quality improvements and reliability improvements, such as downtime reduction and/or extended equipment life, may be motivations for including additional equipment as well. Note that these motivations may also suggest the need for additional inspection, testing, and quality assurance (QA) tasks.

At some facilities, personnel have found that they can improve their credibility with others by not trying to be overly ambitious in their equipment selection efforts. This credibility is important for gaining cooperation in achieving compliance and progressing toward further objectives. Successful MI programs are often implemented with a relatively small initial scope (e.g., as a pilot test) that can be expanded or intensified after MI becomes part of the facility's culture. However, following through with program expansion when practical is important.

## 3.2 ESTABLISHING EQUIPMENT SELECTION CRITERIA

Defining the types of equipment to include may involve interpretation of regulations, clarification of other program objectives, statement of performance objectives, and often some case-by-case decision making. While portions of this exercise may involve uncertainties, establishing and documenting clear criteria that explain the rationale for including and excluding equipment in the program is important to the program's success. (Note: Many times the rationale for excluding equipment is more useful than the rationale for including equipment.) Clearly defining criteria may not prevent equipment selection decisions from being challenged (e.g., by internal or third-party auditors); however, consistent application of clear equipment selection guidelines is more readily defensible than inconsistent application of vague guidelines, even when the clear guidelines are less ambitious. The selection criteria, the process used to define an equipment list using the criteria, and the appropriate roles and responsibilities should be thoroughly documented (see Section 3.4). A simple example of equipment criteria documentation is provided in Appendix 3A.

Examples of types of equipment to include in the MI program are discussed in Chapter 9. This section identifies rationale for establishing equipment selection criteria. Some companies designate safety and/or environmentally critical equipment. Because a common objective within the chemical process industries (CPI) is to contain hazardous chemicals, MI programs almost always include (1) pressure vessels; (2) atmospheric and low-pressure storage tanks; (3) piping and piping components, including valves, in-line filters, eductors, and venturis; and (4) pressure relief devices (e.g., pressure safety valves [PSVs], rupture disks, pressure-vacuum valves, weighted hatches) and the vent systems that protect those items. Listing pressure vessels, tanks, and relief devices is usually straightforward. Adding piping to the equipment list can be more difficult; this will be addressed in the "determining level of detail" discussion in the next section.

## 3. EQUIPMENT SELECTION 19

In addition to pressure vessels, tanks, and piping, facility personnel should consider whether secondary containment components (e.g., dikes, curbs, sumps, other waste collection systems) are to be included. Frequently, environmental and safety (and regulatory) considerations drive companies to include secondary containment. Also, fireproofing on structural components and insulation on tanks (particularly when the relief design takes credit for insulation) should be considered for inclusion in the MI program.

Rotating equipment that contains hazardous fluids needs to be included to help ensure containment of those fluids; however, the following items regarding rotating equipment also should be considered:

- Is ensuring process flow an objective? If so, the drivers (e.g., turbines, motors) may need to be included.
- Which noncontainment items (e.g., agitators, conveyors, fans) are to be included? Recognize the hazards to the process and/or personnel safety when considering loss of function of these items.
- Are nonprocess items (e.g., cooling water systems, steam systems, refrigeration systems, power distribution systems) to be included? If so, is containment of those systems (as well as functionality) considered important? Again, consider the hazards of loss of those systems.

Functional piping components such as filters, eductors, and venturis may similarly have performance objectives.

Instrumentation is an important consideration in most MI programs. Identifying which instruments to include and determining the particular activities desired (e.g., functional tests, QA verifications) can be problematic. Usually, process instrumentation prevents, detects, and/or mitigates unplanned events. Only those instruments that have an impact on potential events related to the MI program objectives need to be considered within the scope of the MI program. Furthermore, for a particular unplanned event with relatively high risks, it may be necessary to ensure more layers of protection.

A variety of methods exist for developing instrumentation lists for MI programs (sometimes called "critical instrument" lists). Some facilities include all instrumentation associated with any other piece of covered equipment; some have asked the Operations department(s) for instrument lists; while others have extracted instrument lists from the safeguards documented in process hazard analysis (PHA) reports. Any of these approaches can work, but each also has typical flaws associated with it. Including all instrumentation that is associated with other covered equipment can be a simple approach to setting up the program, but can result in (1) a long list of instrument testing on the backlog and/or (2) the diversion of resources from more critical MI tasks. Getting input from Operations is beneficial, but without guidelines and examples to follow, operations-generated instrument lists are very inconsistent and hard to defend. Similarly, PHA teams can be an excellent resource; however, using only PHA reports can be inadequate. To increase the effectiveness of the PHA as a resource, PHA teams should be provided with a specific objective of identifying instrumentation to be included in the MI program, along with guidelines for instrument selection and examples.

Using the results of studies conducted to satisfy Instrumentation, Systems, and Automation Society (ISA)/American National Standards Institute (ANSI) S84.01 for safety instrumented systems (SISs) can help facility personnel establish criteria for selecting instrumentation that provides emergency functions. In addition, when layer of protection analysis (LOPA) has been used to analyze facility safeguards, the LOPA results can identify instrumentation (in addition to other safeguards) that is important to process safety. More information on LOPA is provided in Chapter 11.

Utility systems may be critical to process operations and should be considered for inclusion in the MI program. Often, the functionality of the utility is a greater concern than containment. Consider system-wide hazards such as loss of nitrogen, which may lead to explosive vapor spaces in storage tanks. Then consider unique hazards such as loss of containment of the nitrogen system in the analyzer building, which could lead to an asphyxiation hazard. In addition, evaluate the functions of devices such as electrical uninterruptible power supplies (UPSs), emergency communications systems, and electrical grounding and bonding systems. Quality PHA reports should have identified these types of issues, which in turn can be useful for identifying systems/equipment to include in the MI program.

Equipment and systems that are in place to mitigate or to act as the final means to prevent chemical releases, fires, and other catastrophic events should be included in the MI program. This includes fixed and portable fire protection equipment and may also include emergency inerting or "kill" systems (e.g., reaction quenching, reaction mitigation), containment systems, and release detectors.

Another area to consider is equipment that belongs to another company (e.g., chemical vendor, bulk gas supplier) that is connected to a process within the host facility. The host facility will ultimately be accountable for any mishaps that lead to safety, environmental, or plant performance issues; however, the company owning the equipment is generally responsible for equipment maintenance (depending on contractual terms). The vendor's equipment should be evaluated for inclusion in the facility's MI program in the same manner that the facility is evaluating its own equipment. Frequently, the vendor will still perform the MI activities; however, the host facility should take steps to ensure that the vendor's MI activities comply with or exceed the MI program requirements of the host facility.

Likewise, transportation equipment, including that used for onsite storage, is usually considered part of a chemical process during the time it is connected to a process and any time that it is not connected to its motive force (e.g., a truck trailer staged by the process and disconnected from the truck). The MI activities associated with the transportation equipment are frequently the responsibility of the transport company and may be difficult for facilities to track. Facilities should ensure that transport companies are aware of MI requirements and should make provisions to verify that MI activities are occurring. Note that when transport equipment is used for onsite storage connected to a process, it is generally considered part of that process.

# 3. EQUIPMENT SELECTION

The MI program should have provisions for addressing temporary components, such as flexible transfer hoses connecting leased equipment or transportation equipment to a process. In addition, temporary repairs such as leak repair clamps should be addressed (e.g., in the facility's management of change [MOC] program) and are generally not included on equipment lists, but addressed on a case-by-case basis.

MI program developers at some facilities may be inclined to omit traditional safety, fire protection, emergency response, plant evacuation alarms, building ventilation, and/or power distribution equipment from MI equipment lists because this equipment is procured, inspected, and tested by personnel from other departments (e.g., safety personnel, the fire chief) or by contract personnel supervised by other departments. However, MI activities for that equipment must also meet or exceed the MI program requirements, and such MI activities should be appropriately documented.

Finally, consider whether structural components, such as foundations and structural supports (e.g., piping support columns and racks), should be included (separately, as part of associated equipment, or not at all) in the MI program. Consider the age and history of the facility (also consider inspecting new installations for structural defects), the apparent condition of the structural components, and geographical issues (such as the potential for earthquake or hurricane damage) to determine whether to include structural components in the facility's MI program.

## 3.3 DEFINING LEVEL OF DETAIL

Once the equipment selection criteria have been established, facilities can generate an equipment list. Itemizing the equipment can be done at different levels of detail; therefore, facilities should determine an approach for these and similar issues to help ensure consistency:

- Pressure vessels. Generally, facilities include internal coils, liners, and jackets with the pressure vessel. Determine whether to give special designations to any or all of those components.
- Rotating equipment packages. Lube oil systems, seal flush systems, and other support items for large rotating equipment pieces may have piping, pumps, pressure vessels, and instrumentation to include in the MI program. Sometimes, these items are installed together on a skid and without individual equipment numbers. Some facilities assign numbers and list the items separately. Other facilities group the support items with the rotating equipment items.
- Utility systems. Some facilities group an entire system together. Other facilities list the individual system components. Ensure that entire system listings include a full description of the components in the system.
- Vendor packages. Some facilities itemize the separate pieces of vendor equipment. Others group vendor equipment together system by system.

Again, when listing as a system, include a full description of system components.

- Piping. In a relatively simple facility, piping can be listed by system description. More often, facilities use a line numbering system. Whichever system is used, the facility needs to determine how pipe components will be included. Piping appurtenances (e.g., expansion loops, expansion joints, sight glasses, drain lines, nipples, emergency isolation manual block valves, electrical bonding on piping, and cathodic protection) should be appropriately considered. Developing isometric circuit diagrams and/or highlighted piping and instrumentation diagrams (P&IDs) to accompany piping system descriptions is beneficial.
- Instrument loops. Because MI activities are performed by different groups and because the frequency of testing the components may vary, some facilities prefer to list the components of instrument loops separately. Other facilities identify and list the instrument loops themselves. Either approach can be used. Ensure that the functionality and associated logic are tested for the loops even when the equipment list identifies the components separately.
- Pressure relief/vent systems and devices. Many relief devices are individually numbered. However, miscellaneous equipment such as seal loops and weighted hatches are sometimes included as part of the vessel. Ensure that MI tasks are in place for these items. Also, relief device discharge piping is frequently inspected visually as part of relief device servicing. This piping is generally catalogued with the devices. Ensure that the integral components of these systems (e.g., flares, relief catch pots, relief dump tanks, emergency scrubbers) are recognized.
- Fire and chemical release mitigation systems. Similar to the instrument loop considerations, these systems can be treated as systems or individual components. Also, associated piping systems may be numbered or named. Ensure that functional testing verifies mitigation system performance.

The considerations listed above may not be all inclusive. Any inconsistencies in the application of the MI program equipment selection criteria should be revisited and the situation should be corrected.

## 3.4 DOCUMENTING THE EQUIPMENT SELECTION

To ensure that the scope is clearly communicated and understood, the equipment to be included in the MI program should be documented. This document should include the equipment identifier (e.g., tag number) and the equipment name, and may include other comments related to the MI program. The rationale for inclusion and/or exclusion of equipment from the MI program (i.e., the equipment selection criteria) should also be documented to help preserve corporate memory. Thorough documentation will also ensure that auditors are consistently provided a defensible case for excluding an item.

# 3. EQUIPMENT SELECTION

Just as other important documents, the equipment list should be kept current. Equipment additions, deletions, and significant modifications should be tracked via MOC or equivalent programs, and the equipment list should be included as one of the documents that may need to be changed. In addition, other documents that are based on the equipment list (e.g., inspection plans, QA plans) may need to be updated when changes occur.

## 3.5 EQUIPMENT SELECTION ROLES AND RESPONSIBILITIES

The roles and responsibilities for equipment selection can be assigned to personnel in various departments. Typically, personnel in the Maintenance, Engineering, Safety, Health, and Environmental (SHE), and Operations departments will be involved in equipment selection, perhaps as part of a multidiscipline team. Example roles and responsibilities for the equipment selection process are shown in Table 3-1. While the specific information in Table 3-1 may differ from one facility to the next, what remains consistent is the importance of assigning and communicating roles and responsibilities to the appropriate personnel. The matrix designates assignments to personnel with "R" as the person(s) responsible for the activity, "A" as the approver of the work or decisions made by the responsible party, "S" as persons supporting the responsible party in completing the activity, and "I" as persons informed when the activity is completed or delayed.

**24** GUIDELINES FOR MECHANICAL INTEGRITY SYSTEMS

**TABLE 3-1**
**Example Roles and Responsibilities Matrix for Equipment Selection**

| Activity | Maintenance Department Personnel ||| Production Department Personnel ||| Other Personnel ||
|---|---|---|---|---|---|---|---|---|
| | Maintenance Manager | Maintenance Engineers | Maintenance Supervisors | Area Superintendent | Production/Process Engineers | Production Supervisors | PSM/MI Coordinator | PHA Teams |
| **Establishing Selection Criteria** | | | | | | | | |
| • Review program objectives | A | R | R | — | R | — | S | — |
| • Establish rules to follow | A | R | S | — | S | — | S | S |
| • Agree on level of detail | A | R | S | — | S | — | S | — |
| • Document selection criteria | — | R | — | — | R | — | S | — |
| **Generating Equipment List** | | | | | | | | |
| • Apply criteria to equipment lists and/or drawings | | R | S | — | R | — | S | S |
| **Maintaining Equipment List** | | | | | | | | |
| • Review equipment additions, deletions, and modifications | | R | S | — | S | — | S | S |

## 3. EQUIPMENT SELECTION 25

**Appendix 3A. Sample Guidelines for Selecting Equipment for the MI Program.** This appendix documents the criteria that XYZ Chemical Company has established for selecting covered equipment for its MI program. Of particular interest is the nonpressure containing equipment (e.g., instruments) included in the program. To start with, all chemical-containing equipment (e.g., vessels, tanks, piping, pumps) and relief devices are in the program. Further prioritization (i.e., on a system-by-system basis) will be used to designate the criticality of some chemical-containing equipment, which may result in the exclusion of equipment with nonconsequential failures.

Emergency shutdown (ESD) systems are all covered by the XYZ MI program. For the purpose of establishing a covered equipment list, an ESD is limited to shutdown devices that can be actuated manually and remotely to isolate the process. These systems are generally actuated by control room switches. No automatic interlocks are included without meeting the other criteria set below, even if these interlocks result in the shutdown of the process and/or the closing of ESD valves.

In general, the following philosophy is being applied to all other equipment at XYZ:

- All systems that are used for detecting and/or mitigating releases after they have occurred (e.g., area hydrocarbon detectors, deluge systems, dike areas) are included.
- Any device that can detect a deviation that will directly lead to a release (e.g., high-high pressure alarm in a vessel with a relief valve that discharges to atmosphere), and the failure of that device is unannounced, is included in the program.
- Devices that detect process deviations that may lead to other process deviations (e.g., high temperature alarms or high level alarms on vessels for which those deviations may cause high pressure) are not included if the last process deviation before the release (i.e., high pressure for the example above) can be detected.
- When no detection device is available for the last process deviation before a release, then devices detecting the contributing causes are to be included (e.g., safeguards designed to prevent compressor failures or seal failures on single seal pumps).

The following specific scenarios are the results of reviews with XYZ personnel:

- Flare system failures. MI activities that exist for ensuring at least one level of defense against flameouts and loss of purge gas are included in the MI program.
- Liquid in the flare system. During upsets from some processes, liquid hydrocarbon can be released from the flare stack. Therefore, instruments that detect deviations that could contribute to those events (e.g., high level alarms for flare system knockout drums) are covered. For the other process

areas, liquid hydrocarbon is unlikely to be released from the flare stacks and level alarms are not covered.
- Relief devices. All high-pressure alarms/switches on equipment that relieves to the atmosphere are covered. This includes equipment for which the primary relief route is the flare, but a secondary route is to the atmosphere. If no high-pressure detection is available, devices that prevent or detect contributing causes of high pressure are covered.
- Restriction orifices intended to limit relief device release requirements or restrict flows for another safety reason (e.g., safe maximum rate of feed to a reactor) are covered. Note that in many services, corrosion and/or erosion of these orifices is unlikely.
- Pump seal ruptures (single seal hydrocarbon pumps). Devices that are intended to prevent deviations contributing to pump seal failures are covered.
- Pump seal ruptures (double seal pumps). Devices that will alert personnel to the loss of the primary seal (e.g., pressure alarms, sealant system level alarms) are covered. If no devices exist (or if only local indicators are installed), these pumps are to be treated as single seal pumps.
- Direct releases prevented by process interlocks. Interlocks that prevent inadvertent releases of process material directly to the atmosphere (e.g., opening of reactor block valves and dump tank drain valves) are covered.
- Reverse flow of process materials into utility systems or to the atmosphere. Check valves, backflow preventers, and pressure regulators that provide the primary defense against credible, hazardous reverse flow scenarios are to be included in the MI program.
- Dust explosions and/or flammable concentrations of volatile hydrocarbon in finishing areas. Safeguards protecting against these scenarios include nitrogen blanketing in several plants and air sweeping through the finishing system. Devices that detect the loss of such systems (e.g., pressure alarms, flow alarms, analyzers) are to be included in the MI program. XYZ also relies on administrative controls (e.g., laboratory analyses) to prevent/detect flammable concentration scenarios.
- Polymer formation in monomer vessels. Systems and instrument loops (e.g., nitrogen purges) that are used to prevent plugging of nozzles leading to pressure relief devices are included.
- Releases associated with barge and railcar loading/unloading. Hose integrity is to be verified on a periodic basis.
- Overfilling storage tanks (that do not have relief devices). Instrumentation used to help prevent the overflow of hazardous materials is included in the MI program.
- Process motor failures. In general, motors and other drivers will only be included in the MI program if the functioning of the driver is critical to process safety (no specific drivers were identified). Note: Many motors will be included in inspection programs for reliability reasons.

## 3. EQUIPMENT SELECTION

- Building ventilation systems that are used to control the concentration of hazardous vapors and dusts, such as flammable, explosive, or toxic materials, are included in the MI program. In addition, features of buildings themselves (e.g., seals to help ensure positive pressure) should be included in the MI program, as necessary.
- Thermal reactions and chemical decompositions leading to vessel temperatures higher than vessel temperature ratings. Interlocks and alarms in these systems that help detect and/or prevent these events are included in the MI program.
- Freezing of water and other chemicals in process equipment. Generally, heat-tracing systems designed for freeze protection are included in the MI program.
- Cooling systems, including tracing systems, that are used to help prevent runaway reactions (e.g., at monomer storage tanks, in services with heat sensitive materials) are included in the MI program.
- Compressor failures. Systems and instrument loops that are used to protect compressors (e.g., high- and low-pressure instrumentation, high-temperature instrumentation, high liquid level instruments in accumulators, vibration instruments, lube oil systems and related instruments, intercoolers, knockout drums, and liquid removal systems) are included in the MI program.
- Releases of process materials to the sewer or liquid to the flare system as a result of low interface levels. Interface level instrumentation related to these scenarios is to be included in the MI program.
- Known scenarios (e.g., polymer formation) for plugging flare lines. Systems and instrument loops used to prevent blockage of pressure relief device vent headers (e.g., process vessel liquid level instrumentation, nitrogen purge systems, liquid drain systems, vent header pressure monitoring instrumentation) are included in the MI program.
- Failures of rupture disks beneath relief valves. Systems and instrumentation used to detect leaking or burst rupture disks are included in the MI program.
- Reactor upsets. Systems and instrumentation used to prevent runaway reactions (e.g., inhibitor or poison injection systems, dump systems, quench systems, pressure venting systems) are included in the MI program.
- High level in vessels (e.g., seal pots) downstream of relief valves. Systems and instrumentation that are used to maintain or monitor the liquid level in seal pots, as well as systems and instrumentation that are used to prevent overfilling of seal pots, are to be included in the MI program.
- Utility failures. Safeguards on utility systems, including low-pressure alarms, air dryers, emergency lighting panels, and indicators that confirm that backup systems are operating, are included in the MI program. Emergency utility systems (e.g., electric power supplies, instrument air supplies, emergency cooling systems) are to be included in the MI program. All of the parts and instrumentation for each emergency system are to be included.

- Control valves at high pressure/low pressure interface points. The integrity of the valve, as well as the logical actions governing valve position, are included in the MI program.
- Heat exchangers. When tube leaks will result in adverse consequences, the exchanger tubes are included in the MI program.
- Systems that are used to enhance the reliability of covered instrument loops, such as chemical seals and instrument cabinet gas purges, are included in the MI program.
- Miscellaneous other equipment to consider for the MI program:
    - Blast walls, bunkers, barricades, and blast-resistant doors for control rooms and other occupied process buildings
    - Barriers used to protect process equipment from vehicles
    - Pipe galleries carrying process piping, including spring hangers and other supports
    - Excess flow valves
    - Structural steel
    - Fire protection insulation applied to structural steel and to process vessels
    - Instrumentation designed to detect heat, flammable vapors, or toxic vapors
    - Vent handling systems such as flares, scrubbers, and thermal oxidizers
    - Emergency response equipment, including ambulances, fire trucks, self-contained breathing apparatus (SCBA), radios, and rescue equipment
    - Portable oxygen and flammable gas meters
    - Testing and inspection test equipment
    - Cranes and other lifting equipment
    - Tank internals (e.g., heating or cooling coils, agitator baffles, steady bearings, roof supports)
    - Process sewers, including liquid seals, purge systems, and sewer ventilation systems
    - Valves in pressure relief lines, including three-way valve assemblies and drains
    - Weather protection bags or socks placed over vertical vent lines from pressure relief devices
    - Liquid expansion bottles, including rupture disks and heat tracing, on lines that do not have pressure relief devices
    - Inerting systems and associated instrumentation
    - UPS systems for instrumentation, area alarms, public address systems, and communication systems
    - High-temperature monitor and interlock instrumentation for solids processing equipment that could have potential hot spots, such as bearings
    - Electrical grounding and bonding systems

# 4
# INSPECTION, TESTING, AND PREVENTIVE MAINTENANCE

Once the scope of the mechanical integrity (MI) program has been defined, MI program efforts often focus on developing and implementing the inspection, testing, and preventive maintenance (ITPM) program. (Note: The use of the term "preventive maintenance" in ITPM includes all proactive maintenance tasks that are not considered to be inspection or test tasks, and are performed to prevent or predict failure of equipment included in the MI program. Also note that preventive maintenance [PM] is not intended to imply that every task traditionally viewed as a PM task is included in the MI program. Only PM tasks needed to prevent failures of interest [e.g., loss of containment of hazardous materials are included in an ITPM program.) In many respects, the ITPM program is the core of the MI program. The ITPM program's objective is to identify and implement maintenance tasks needed to ensure the ongoing integrity of the equipment. Doing so provides a facility with the opportunity to move away from the "breakdown" maintenance philosophy (Reference 4-1) and move to a more proactive philosophy of maintaining equipment integrity. For the purpose of this chapter, ITPM program development and implementation involve the following two phases:

1. ITPM task planning. The activities of this phase include identifying and documenting the ITPM tasks needed to ensure ongoing integrity of equipment and to establish the frequency to perform those tasks. Next, the tasks are translated into a schedule. (Note: An ITPM task is an inspection, testing, or preventive activity that is performed at some interval with the purpose of ensuring the ongoing integrity of equipment.)
2. ITPM task execution and monitoring. The plan becomes reality when the tasks are executed by qualified personnel as scheduled. To help achieve this objective, the ITPM program needs to ensure that processes are established to monitor the schedule, the task results, and overall program performance.

The result of these activities is a proactive maintenance program that helps ensure equipment integrity. In addition, continuous monitoring of the task schedule and results provides a means for optimizing the tasks.

Many organizations find that the ITPM program includes maintenance management activities (e.g., developing and managing the ITPM schedule) and proactive maintenance tasks that are similar to those in reliability programs. An MI program often focuses on equipment integrity issues necessary for ensuring safety and protecting the environment. A reliability program similarly focuses on equipment integrity issues needed to ensure business performance (e.g., reducing process downtime). Used together, these programs implement maintenance management activities and proactive maintenance tasks to detect, prevent, and manage equipment failures before they impact performance. Therefore, concepts and processes used to develop and implement the ITPM program (and the entire MI program) also can be used to implement and/or improve a reliability program.

## 4.1 ITPM TASK PLANNING

To develop the ITPM program, facility personnel often identify and document the ITPM tasks and then develop a schedule for each of these recurring tasks. This book refers to the list of recurring tasks and their associated schedule as the ITPM plan. The key attributes of this plan are that it:

- Includes recurring ITPM tasks for all equipment within the scope of the MI program
- Defines the basis for each ITPM task and its corresponding interval
- Provides a link to procedures and other necessary references (e.g., original equipment manufacturer [OEM] manuals)
- Defines (or references the location of) acceptance criteria for each ITPM task

This chapter outlines the following activities and considerations for developing an ITPM plan that meets those attributes (Reference 4-2):

- Selecting ITPM tasks
- Developing sampling criteria
- Other ITPM planning considerations
- Determining ITPM task scheduling

The following subsections discuss each of these areas in more detail.

### 4.1.1   ITPM Task Selection

The starting point in developing the ITPM plan is selecting the recurring ITPM tasks. Each equipment item within the MI scope must be included in the task selection process. A five-step process for selecting ITPM tasks is illustrated in Figure 4-1 and described in detail in the following paragraphs.

## 4. INSPECTION, TESTING, AND PREVENTIVE MAINTENANCE

*Step 1 – Organize Equipment into Equipment Classes.* While ITPM task selection can be performed for each individual equipment item, grouping items into equipment classes (e.g., pressure vessels, centrifugal pumps) can reduce the time needed to select tasks and help ensure program consistency. The concept behind equipment classes is that the ITPM tasks selected are applicable to all equipment items within the class. However, when developing equipment classes, personnel must ensure that unique items and/or items in different service conditions (e.g., different chemicals, higher pressures) are identified and separated into sub-classes when different ITPM tasks and intervals are warranted. Furthermore, special cases, such as particularly problematic equipment, may merit unique treatment in the ITPM plan.

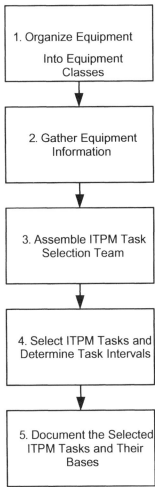

**FIGURE 4-1** ITPM task selection process.

Organizing the equipment into classes provides additional advantages:

- Potential for fewer procedures needed. Some regulations require procedures for every ITPM task. Using the equipment class approach, the procedures can typically be developed for each task by equipment class rather than for each item within the equipment class. Chapter 6 discusses the identification of procedure needs in more detail.
- Improved identification and communication of responsibilities for ITPM tasks. The equipment class-based ITPM plan typically documents responsibilities for tasks for each equipment class.
- Consistent selection of ITPM tasks and intervals for added equipment. As processes and equipment are added, an equipment class-based ITPM plan makes it easier to define the ITPM tasks and intervals for the additions.
- Efficient ITPM program documentation for regulators and auditors. Early in an audit of an MI program, the regulator/auditor typically requests information on the ITPM tasks and their intervals. The equipment class-based ITPM plan provides that documentation concisely.

***Step 2 – Gather Equipment Information.*** To select the ITPM tasks and their intervals efficiently, the following types of information about the equipment and its operation should be compiled prior to task selection:

- Engineering data, such as design specifications and as-built drawings.
- Operational data, such as procedures that contain operating parameters and operating limits tables.
- Maintenance and inspection history, including current ITPM tasks and schedule, and inspection and repair history.
- Safety and reliability analyses (e.g., process hazards analyses, reliability-centered maintenance [RCM] analyses) that can provide information about types of failures expected and the effects of those failures. Such analyses may also identify important protective equipment (e.g., alarms, interlocks, emergency response, critical utilities) whose functionality needs to be maintained.
- Applicable recognized and generally accepted good engineering practice(s) (RAGAGEPs) for equipment, especially those that specify ITPM tasks and intervals.
- Applicable regulatory/jurisdictional requirements.
- Applicable site or corporate environmental, health, and safety policies.
- OEM manuals.
- Risk-based analyses, if the ITPM task selection and/or interval determination is to be risk-based. Examples of such analyses include quantitative risk assessments, layer of protection analyses (LOPAs), and risk-based inspection (RBI) studies. Step 4 of the ITPM task selection process and Chapter 11 include additional risk-based ITPM task selection processes and applicable risk analysis techniques.

# 4. INSPECTION, TESTING, AND PREVENTIVE MAINTENANCE

Some of the above information is equipment-specific and usually kept in the facility's equipment files. Table 4-1 provides a list of information to include in equipment files for selected types of equipment in order to provide the proper information needed for the MI program, including the ITPM program. Equipment information can be useful for selecting tasks by:

- Identifying the RAGAGEPs to be considered
- Providing information on potential or known damage mechanisms
- Identifying candidate ITPM tasks
- Providing regulatory/jurisdictional and/or site or corporate ITPM requirements

In addition, equipment information can assist in the following ITPM activities:

- Defining acceptance criteria, usually by referencing specific equipment file information (e.g., corrosion allowance) (see Section 4.2.1 for additional information on acceptance criteria)
- Preparing for execution of an ITPM task (see Section 4.2.2 for additional information on the need for equipment information during ITPM task execution)

***Step 3 – Assemble ITPM Task Selection Team.*** With the breadth of knowledge needed to select tasks, employing a multi-discipline team is desirable. Typical personnel included on the team are:

- Engineering personnel, who can provide knowledge of the equipment design and applicable codes, standards, and recommended practices.
- Operations personnel, who can provide knowledge of the equipment operation and failure history. In addition, having operations involved in the task selection process helps promote buy-in to the ITPM plan, which can pay dividends during the task execution.
- Maintenance personnel, who can provide knowledge of current maintenance practices and maintenance history.
- Inspection personnel, who can provide knowledge of inspection and testing codes, standards, recommended practices, potential damage mechanisms, and inspection history.
- Reliability and maintenance engineers, who can provide knowledge of inspection, PM, potential damage mechanisms, and equipment history.
- Corrosion engineers, who can provide knowledge of corrosion and other damage mechanisms (e.g., stress-induced cracking) and corrosion prevention and monitoring techniques.
- Process engineers, who can provide knowledge of equipment design and operation; equipment history; and applicable codes, standards, and recommended practices.

**TABLE 4-1**
**Typical Equipment File Information for Selected Types of Equipment**

| Equipment Type | Design and Construction Information (Ref. 4-3) | Service History | ITPM History | Vendor-supplied Information |
|---|---|---|---|---|
| Pressure Vessels and Atmospheric Storage Tanks | • Design specification<br>• Design code<br>• Materials of construction<br>• Corrosion allowance | • Fluids handled<br>• Type of service (e.g., continuous, intermittent, irregular)<br>• Vessel/tank history (e.g., alterations, date of repairs)<br>• Operating parameters<br>• Temperature/pressure excursions<br>• Failures and repair history | • Inspections performed<br>• Examination techniques used<br>• Inspector qualifications<br>• Inspection results | • Data reports (e.g., U-1 form, API 650 form)<br>• Type of construction<br>• Rubbings/photocopies of code nameplates<br>• As-built drawing |
| Piping | • Piping specification<br>• Design code<br>• Corrosion allowance<br>• Welder qualifications<br>• Piping and instrumentation diagrams (P&IDs) and process flow diagrams (PFDs) | • Fluids handled<br>• Type of service (e.g., continuous, intermittent, irregular)<br>• Operating parameters<br>• Temperature/pressure excursions<br>• Failures and repair history | • Circuit definition and thickness measurement locations (TMLs), usually on isometric drawings<br>• Inspections performed<br>• Examination techniques used<br>• Inspector qualifications<br>• Inspection results | • Equipment manuals for system components (e.g., filters, valves) |
| Relief Devices | • Design specification<br>• Relief design basis and calculations<br>• Materials of construction<br>• P&IDs and PFDs | • Fluids handled<br>• Failures and repair history | • Inspections and tests performed<br>• Inspection and test results<br>• Testing organization certification/qualifications | • Equipment manuals<br>• Manufacturer's data report<br>• Rubbings/photocopies of code nameplates |

## 4. INSPECTION, TESTING, AND PREVENTIVE MAINTENANCE

**TABLE 4-1 (Continued)**

| Equipment Type | Design and Construction Information (Ref. 4-3) | Service History | ITPM History | Vendor-supplied Information |
|---|---|---|---|---|
| Rotating Equipment | • Equipment specification<br>• Materials of construction<br>• Seal configuration and data | • Fluids handled<br>• Type of service (e.g., continuous, intermittent, irregular)<br>• Type of lubricant<br>• Operating parameters<br>• Temperature/pressure excursions<br>• Failures and repair history | • Inspections and PM performed<br>• Inspection and PM results | • Equipment manuals<br>• Manufacturer's data report (e.g., API 610 form)<br>• Performance testing data (e.g., pump curves)<br>• Recommended spare parts list<br>• As-built drawing |
| Instrumentation | • Instrument specification<br>• Materials of construction<br>• Line diagrams<br>• Logic diagrams<br>• Required safety integrity level (SIL) and safety requirement specification | • Corrosive or fouling service<br>• Failures and repair history | • Testing performed (e.g., calibration)<br>• Testing results | • Instrument manuals<br>• Factory calibration report |
| Electrical Equipment | • Equipment specification<br>• Line diagrams<br>• Over-current protection information<br>• Logic diagrams (e.g., redundant power supply switching logic) | • Failures and repair history | • Testing performed (e.g., infrared analysis, transformer oil analysis)<br>• Testing results | • Equipment manuals<br>• As-built drawings |

- Inspection and maintenance contractors. If facility management plans to use outside contractors for inspection and nondestructive testing (NDT) tasks, including contractor representatives on the team can be beneficial.
- Equipment manufacturers and vendor subject matter experts (SMEs). These individuals are especially helpful when selecting ITPM tasks for licensed processes and new equipment because they can provide valuable knowledge about the operation and maintenance of processes and equipment when facility personnel lack pertinent experience.

***Step 4 – Select ITPM Tasks and Determine Task Intervals.*** When selecting the ITPM tasks, the team should consider the types of failures to be addressed (e.g., general corrosion, unavailability of an instrumented shutdown system) and the best approach (and most effective tasks) for detecting or preventing the failure. The team may have many failures to consider; however, the task selection process can be simplified by understanding that for most processes that use hazardous chemicals, the failure modes that most need to be addressed relate to:

- Preventing loss of containment
- Preventing or detecting the loss of functionality of critical controls, safety systems (e.g., alarms, interlocks), emergency response equipment, and critical utilities
- Preventing unnecessary shutdown and startup of processes in which shutdown and/or startup introduces a hazardous condition in the process

In addition, identification of specific failure/damage mechanisms (e.g., localized corrosion, erosion, signal drift) is needed to ensure that proper ITPM tasks and intervals are selected. This is especially true of pressure-boundary equipment, such as pressure vessels, storage tanks, process vessels, piping, and so forth. The following is a partial list of general failure/damage mechanisms (Reference 4-4) for pressure-boundary equipment:

- Mechanical loading failures, such as ductile fracture, brittle fracture, mechanical fatigue, and buckling
- Wear, such as abrasive wear, adhesive wear, and fretting
- Corrosion, such as uniform corrosion, localized corrosion, and pitting
- Thermal-related failures, such as creep, metallurgical transformation, and thermal fatigue
- Cracking, such as stress-corrosion cracking
- Embrittlement

The American Petroleum Institute (API) Recommended Practice (RP) 571, *Damage Mechanisms Affecting Fixed Equipment in the Refining Industry*, provides additional information on damage mechanisms for fixed equipment.

The team should determine the best approach(es) for managing each identified failure. In general, failures can be managed by implementing ITPM tasks that:

## 4. INSPECTION, TESTING, AND PREVENTIVE MAINTENANCE 37

- Detect the onset of a failure condition (e.g., presence of a crack, excessive pump vibration)
- Assess the condition of equipment (e.g., remaining life based on corrosion rate, accuracy of instrumentation)
- Prevent premature failure (e.g., replacement of worn compressor parts, lubrication of pumps)
- Discover hidden failures (e.g., verifying the operation of emergency shutdown [ESD] systems, confirming the operation of interlocks, verifying the operation of backup power supply systems)

After identifying the desired approach(es) for managing each failure, the team can then determine the type(s) of task(s) needed. In general, the tasks fall into the following three categories:

1. Inspection tasks, which detect the onset of a failure condition (e.g., vessel wall crack) and/or assess the condition of the piece of equipment (e.g., vessel wall thickness).
2. Testing tasks, including predictive maintenance tasks, which assess the condition of the equipment (e.g., drift in an instrument, vibration of a pump) and/or detect hidden failures (e.g., functional test of a shutdown system).
3. PM tasks, which help prevent premature failure of the equipment by (1) promoting the equipment's inherent reliability (e.g., lubricating a pump) or (2) restoring the equipment's reliability by replacing the entire item or selected components/parts (e.g., rebuilding of a compressor before its functionality is lost).

Once the desired type(s) of task(s) (i.e., inspection, testing, and/or PM) is identified, the team can employ several resources to assist in selecting specific task(s). In general, the following resources can be used:

- Inspection codes, standards, and recommended practices
- Manufacturer's recommendations
- Professional organization or trade group guidance
- Insurance company recommendations
- In-house history of the equipment under consideration
- Common industry practices (e.g., vibration analysis of rotating equipment)
- Environmental, health, and safety recommendations
- Risk analysis recommendations

Appendix 4A, at the end of this chapter, contains a brief description of some common predictive maintenance and NDT tasks. In addition, API RP 571 includes information on common damage mechanisms, and inspection and monitoring techniques for many types of fixed process equipment (e.g., tanks, piping, vessels). This information can help the ITPM team determine appropriate ITPM tasks.

Team members should ensure that each task selected addresses the types of failures targeted for prevention. Furthermore, team members should verify that any ITPM codes, standards, and recommended practices selected are consistent with the applicable design codes and standards (e.g., API 510 activities are applied to American Society of Mechanical Engineers [ASME]-coded pressure vessels). For some equipment (e.g., specialized equipment and equipment not governed by RAGAGEPs), OEM manuals may list ITPM recommendations. For some equipment, the team will need to decide whether inspection or replacement of an equipment item is preferable. Also, the team may also determine that run-to-failure is an acceptable approach for managing some potential failures. Section 4.1.3 provides additional information on these issues.

To finalize ITPM task selection, the team should decide whether to change the current ITPM tasks and/or frequencies or to add new ITPM tasks if no tasks are in place to address failures of interest. Table 4-2 provides a decision matrix that provides guidelines for this decision.

Once a task is selected, the team should determine the appropriate task interval. (Note: Task interval is the time period between successive task executions.) For some tasks (especially new tasks), this task interval should be viewed as the starting point for establishing the ongoing task interval (if continued performance of the task is needed). Other factors, such as the equipment's operating and maintenance history, often dictate the ongoing interval. These factors vary somewhat for different types of equipment, but in general the factors provide insight into:

- Probability of equipment failing before the next ITPM task
- Probability of the ITPM task detecting the equipment failure mode of interest
- Consequence of the equipment failure

For example, the task interval for inspecting pressure vessels, tanks, and piping can be influenced by:

- Age of the equipment
- Materials handled
- Process conditions (e.g., pressure, temperature)
- Material of construction
- Types of damage mechanisms that the equipment is subject to (e.g., general corrosion, localized corrosion, stress-corrosion cracking)
- Rate of damage/deterioration (e.g., corrosion rate)
- Inspection/maintenance history
- Type of inspection and/or testing technique used, including the effectiveness of the techniques to detect and quantify damage
- Current condition of the equipment
- Codes and/or jurisdictional requirements

## 4. INSPECTION, TESTING, AND PREVENTIVE MAINTENANCE

**TABLE 4-2**
**Example ITPM Task Selection Decision Matrix (Ref. 4-1)**

| | ...and operating history is good and well documented: | ...and there is a history of operating problems: | ...and there is insufficient operating data or documentation available: |
|---|---|---|---|
| If present practices (1) comply with codes, standards, recommended practices, manufacturer's recommendations, etc, and (2) address anticipated failure mechanisms/modes... | • Document the present practice in the ITPM plan and reference applicable code(s), standard(s), RP(s), manufacturer's recommendation(s), etc. | • Review the equipment and operations to ensure that failure mechanisms/modes are adequately addressed by tasks in the ITPM plan.<br>• Consider whether more rigorous ITPM would be appropriate. | • Document the present practice in the ITPM plan and reference applicable code(s), standard(s), RP(s), manufacturer's recommendation(s), etc.<br>• Where possible, identify data to be gathered and documented to validate ITPM task selection. |
| If present practices (1) comply with codes, standards, recommended practices, manufacturer's recommendations, etc, but (2) do not address anticipated failure mechanisms/modes... | • Review the equipment and operations to determine if the unaddressed failure mechanisms/modes are important.<br>• Consider whether more rigorous ITPM would be appropriate. Otherwise, document the present practice in the ITPM plan and reference applicable code(s), standard(s), RP(s), manufacturer's recommendation(s), etc. | • Review the equipment and operations to identify all failure mechanisms/modes not adequately addressed.<br>• Determine ITPM needed to address all failure mechanisms/modes. | • Review the equipment and operations to identify all failure mechanisms/modes not adequately addressed.<br>• Determine ITPM needed to address any failure mechanisms/modes. Otherwise, document the present practice in the ITPM plan and reference applicable code(s), standard(s), RP(s), manufacturer's recommendation(s), etc.<br>• Ensure that steps are taken (e.g., procedures are modified) to improve documentation practices. |
| If present practices (1) do not comply with codes, standards, recommended practices, manufacturer's recommendations, etc, but (2) address anticipated failure mechanisms/modes... | • Document present practice in the ITPM plan and provide justification for any variance from applicable codes, standards, RPs, manufacturer's recommendations, etc.<br>• Ensure that documentation supports your decision. | • Review the equipment and operations to identify all failure mechanisms/modes not adequately addressed.<br>• Determine ITPM needed to address all failure mechanisms/modes. | • Consider upgrading the present practices to meet applicable codes, standards, RPs, manufacturer's recommendations, etc.<br>• Ensure that steps are taken (e.g., procedures are modified) to improve documentation practices. |

**TABLE 4-2 (Continued)**

| | ...and operating history is good and well documented: | ...and there is a history of operating problems: | ...and there is insufficient operating data or documentation available: |
|---|---|---|---|
| If present practices (1) do not comply with codes, standards, recommended practices, manufacturer's recommendations, etc, and (2) do not address anticipated failure mechanisms/modes... | • Review the equipment and operations to determine if unaddressed failure mechanisms/modes are important.<br>• Consider whether more rigorous ITPM would be appropriate. Otherwise, document the present practice in the ITPM plan and reference applicable code(s), standard(s), RP(s), manufacturer's recommendation(s), etc.<br>• Ensure that documentation supports your decisions. | • Review the equipment and operations to identify all failure mechanisms/modes *not* adequately addressed.<br>• Determine ITPM needed to address all failure mechanisms/modes.<br>• Consider upgrading the present practices to meet applicable codes, standards, RPs, manufacturer's recommendations, etc. | • Review the equipment and operations to identify all failure mechanisms/modes *not* adequately addressed.<br>• Consider upgrading the present practices to meet applicable codes, standards, RPs, manufacturer's recommendations, etc. |
| If no accepted codes, standards, recommended practices, manufacturer's recommendations, etc., are available... | • Document the present practice(s) in the ITPM plan and include a "performance history" notation as the rationale for the practice(s). | • Consider the consequences of the equipment failure. If the consequences of failure are hazardous, contact vendors, chemical suppliers, other users, etc., for suggestions. | • Consider the consequences of the equipment failure. If the consequences of failure are hazardous, contact vendors, chemical suppliers, other users, etc., for suggestions. |

## 4. INSPECTION, TESTING, AND PREVENTIVE MAINTENANCE 41

In addition, many of the RAGAGEPs provide information and guidance on these factors (and others) and how they can influence task intervals. Table 4-3 provides some of the factors to consider when establishing task frequencies for relief devices, instrumentation, and rotating equipment.

**TABLE 4-3**
**Factors Affecting ITPM Tasks for Relief Valves, Instrumentation, and Rotating Equipment**

| Relief Devices | Instrumentation | Rotating Equipment |
|---|---|---|
| • Type of device<br>• Relief design case<br>• Process conditions<br>• Susceptibility to plugging<br>• Test/maintenance history | • Type of instrument (e.g., pressure, level)<br>• Vintage of the instrumentation (e.g., pneumatic, electronic)<br>• Measurement technology used (e.g., radar level detection)<br>• Process conditions<br>• Environmental conditions (e.g., outside installation)<br>• Type of testing used<br>• Test/maintenance history<br>• Desired performance (e.g., SIL of a safety instrumented system [SIS]) | • Type of rotating equipment<br>• Age of the equipment<br>• Type of seal system (e.g., mechanical, packing)<br>• Process conditions<br>• Type of ITPM task used<br>• Test/maintenance history |

Performing a task less frequently than recommended by codes, standards, recommended practices, manufacturer's recommendations, and common industry practices can be difficult to justify. The decision to perform a task less frequently will need to be based on the operating experience and the level of documentation available to justify the variance. (Note: Caution should be used when basing task intervals exclusively on operating history because the history may not provide a satisfactory indication of low frequency/high consequence failures.) Again, the decision matrix in Table 4-2 can be applied to this decision.

In addition to the equipment maintained by the facility, the ITPM plan should include equipment owned and maintained by others if the equipment is included in the MI program (e.g., nitrogen storage and supply system). To develop this portion of the ITPM plan, a representative from the equipment owner's organization can be asked to supply the necessary information. The ITPM plan should clearly document which organization is responsible for performing these ITPM tasks (i.e., the facility staff or the equipment owner).

More detailed information on ITPM tasks and intervals for specific types of equipment is provided in Chapter 9 and on the CD accompanying this book. Many companies are using risk-based approaches to determine ITPM tasks and intervals. For example, companies are using (1) RBI assessments to determine the type and frequency of inspection tasks for pressure containing equipment (e.g., vessels, tanks, piping), (2) RCM analyses to determine type and frequency of ITPM tasks for active equipment items (e.g., pumps, controls, compressors),

(3) LOPA and similar analysis techniques used to determine desired performance (safety integrity level), and/or (4) unavailability analysis (e.g., fault tree analysis [FTA], Markov modeling, simplified equations) to determine testing frequency for safety instrumented functions (SIFs) (e.g., shutdowns, interlocks). These approaches all focus on identifying potential loss events, understanding the risks of each loss event, and defining ITPM tasks to effectively manage the risk. Each of these approaches is discussed in more detail in Chapter 11.

*Step 5 – Document the Selected ITPM Tasks and Their Bases.* Finally, the selected tasks should be documented in the ITPM plan. The basis (i.e., rationale) for selecting each task should also be included in the ITPM plan. In general, the basis will be (1) a specific code, standard, or recommended practice, (2) a manufacturer's recommendation, (3) an industry practice, and/or (4) the facility's operating history and/or maintenance practice. Documentation of the plan can be provided in a table format in the facility's MI program document and/or on the facility's computerized maintenance management system (CMMS). Table 4-4 provides an example table format.

### 4.1.2 Developing Sampling Criteria

Another element of ITPM task planning is the application of sampling, especially for inspections, NDT, and condition-monitoring tasks. Sampling can be applied to determine the specific location and number of representative sample points needed for NDT and condition-monitoring tasks. As mentioned earlier, inspection and condition-monitoring tasks focus on detecting failure and/or assessing the condition of equipment. The nature of these tasks and the size of equipment for which they are typically used require that sampling be used to assess the integrity of the equipment.

The location and number of NDT sample points or condition-monitoring locations (Reference 4-5) are determined based upon several factors, such as:

- The consequences of failure. Failures with higher potential for safety or environmental consequences require more sample points.
- The expected deterioration rate (e.g., corrosion rate). In general, cases of high (or higher than expected) rates of deterioration require more sample points.
- The presence of localized deterioration. Higher potential for localized deterioration requires more sample points.
- The equipment configuration. For example, piping circuits with more elbows, tees, and injection points require more sample points.
- The potential for unusual damage mechanisms. In general, equipment with the potential for unusual damage mechanisms (e.g., corrosion under insulation) requires more sample points.

## 4. INSPECTION, TESTING, AND PREVENTIVE MAINTENANCE

**TABLE 4-4**
**Example ITPM Plan in Tabular Format**

| Equipment Class | Required Activity | Interval or Frequency | Basis | Procedure | Assigned Department | Remarks |
|---|---|---|---|---|---|---|
| Pressure vessels | Visual surveillance | Daily | Performance history | MI-1 | Operations | Completed during routine rounds by operators; deficiencies are documented, but surveillance that does not identify a deficiency is not documented |
| | External inspection | 5 years or ½ of remaining life (whichever is less) | API 510 | MI-2 | MI Department | Visual inspection and thickness measurement |
| | Internal inspection | 10 years or ½ of remaining life (whichever is less) | API 510 | MI-2 | MI Department | Visual inspection and thickness measurement |
| Pressure transmitters, including alarms and interlocks | Visual surveillance | Weekly | Performance history | MI-1 | Operations | Completed during routine rounds by operators; differential pressures are monitored if leak in tube is suspected; deficiencies are documented, but surveillance that does not identify a deficiency is not documented |
| | Calibration of transmitters | Annually | Industry practice | MI-3 | Electrical and instrumentation (E&I) Department | |
| | Functional test of alarms and interlocks operation | Bi-annually (every 2 years) | Industry practice | MI-4 | E&I Department | |

RAGAGEPs provide guidance for determining sample number and sample locations. For example, API 570 and API RP 574 provide guidance on location and number of thickness measurement locations (TMLs) for piping inspections (References 4-5 and 4-6). In addition, some RBI approaches include statistical techniques for determining, or accounting for (in the data analysis), the number of sample points.

### 4.1.3 Other ITPM Task Planning Considerations

In addition to task selection and frequency determination, the ITPM task selection team will also have other decisions to make. Specifically, it will need to decide (1) which ITPM tasks are best performed by operations personnel (instead of maintenance personnel), (2) the value of performing ITPM tasks versus periodic equipment replacement, and (3) the appropriateness of allowing certain equipment to run to failure.

*Operator Performed Tasks.* Some ITPM tasks are best performed by operations personnel. Some of these tasks are operator activities that were in place before the ITPM planning effort. These activities include (1) visual surveillance of equipment performed during operator rounds, (2) routine inspection of pump seals for leakage, (3) verifying lubrication fluid levels, (4) listening for unusual sounds, and (5) functional testing of emergency isolation valves that are also used in normal operation. In addition, the ITPM plan may contain additional activities that can be performed by operating personnel. Depending on specific facility culture, these activities may include routine lubrication of equipment, tightening valve packing, and changing oil in selected equipment. Such activities are important elements of the MI program, and the ITPM plan should include these activities and reflect that they are assigned to operations personnel.

*Inspect or Replace?* The ITPM task selection team should also decide whether to perform inspections and tests or to replace an equipment item on a routine basis. Typically, replacement is considered only for relatively inexpensive equipment items (e.g., pressure gauges, small relief valves, small chemical dosing pumps). For some equipment items, this decision requires considering the risk of failure as well as the economics involved. When considering the risk of failure before replacement, the following issues may be important:

- Is the time to failure predictable so that an appropriate replacement time (with little chance of failure) can be determined?
- Is the consequence of equipment failure acceptable?
- If the consequence is not acceptable, do cost-effective inspections and/or tests exist that can help detect the onset of failure?
- Are quality assurance (QA) practices in place to help ensure that the facility is not increasing risks when replacing the equipment (e.g., as a result of improper procurement or improper installation)?

## 4. INSPECTION, TESTING, AND PREVENTIVE MAINTENANCE

Answering these questions does not necessarily require in-depth analysis; however, it is important to consider potential safety and environmental impacts, as well as economics.

***Run-to-Failure?*** Another consideration in ITPM task planning is when run-to-failure can be applied as an appropriate maintenance strategy. Several areas can be addressed to proactively manage failure and to distinguish run-to-failure items from breakdown maintenance items. Specifically, facility management should commit to proactive management of the failure by ensuring that (1) spare parts are obtained, (2) OEM manuals and/or repair procedures are available, and (3) applicable training is conducted to perform the repair.

Deciding to use a run-to-failure maintenance strategy should be done with caution, and the decision should be based on established criteria. The criteria that are used should include:

- No significant safety, environmental, or operational impacts occur as a result of the failure.
- A plan is in place (e.g., a preapproved temporary operating procedure) for continued operation after the failure occurs.
- The cost to repair the failure is lower than the cost of the ITPM tasks needed to prevent the failure.
- The failure priority is too low to warrant resources for ITPM (when compared with resources needed to perform ITPM tasks on higher risk failures).

The responses to the evaluation of these issues should be documented. In addition, facility personnel should also review any supporting process safety documentation (e.g., process hazard analyses [PHAs]) to ensure that these documents are consistent with the run-to-failure decision.

### 4.1.4 ITPM Task Scheduling

Once the ITPM plan is developed, it should be translated into an executable schedule. The objective of the schedule is to logically organize the ITPM tasks into executable work packages by distributing the tasks (over time) based on several factors, including equipment availability and resources. The facility staff should include plans in the MI program for managing and maintaining this schedule.

The basic elements of the schedule are obtained by combining information from the equipment list and the ITPM plan. For example, if the ITPM plan specifies that external visual inspection of pressure vessels takes place every 5 years and the equipment list contains five pressure vessels, the schedule will include an external visual inspection every 5 years for each of the five pressure vessels. This information can then be translated into a time-based schedule.

While many tools (e.g., CMMS) and scheduling techniques are available to help organizations develop the ITPM task schedule, personnel should consider several factors necessary for developing the schedule. Three primary factors for most ITPM tasks are:

- Equipment availability. For some ITPM tasks, the equipment, the process, and/or the entire facility will need to be out of service for a specific time period to execute the task. Often, this requires scheduling out-of-service ITPM tasks during scheduled facility outages (which may need to be adjusted to meet the task frequency in the ITPM plan).
- Personnel availability. Sometimes the schedule is constrained by (1) the number of personnel needed to perform the tasks in the allotted time and/or (2) the availability of personnel with the training and qualifications needed to perform a task.
- Spare part and maintenance material availability. For tasks that involve rebuilding equipment, the availability of spare parts and maintenance materials must also be considered.

For most facilities, the tasks are scheduled by (1) entering them as "recurring work orders" in the CMMS and/or (2) organizing the tasks into groups (e.g., lubrication routes). In addition, the tasks assigned to operations will need to be integrated into the operator work routines and procedures.

## 4.2 TASK EXECUTION AND MONITORING

After establishing the ITPM plan and schedule, personnel should address several factors to help ensure the success of the ITPM program. The following topics are addressed in the subsections below:

- Defining acceptance criteria
- Equipment and ITPM task results documentation
- ITPM task implementation and execution
- ITPM task results management
- Task schedule management
- ITPM program monitoring

Additional considerations include the implementation of personnel training and procedures to support the ITPM plan. These topics are addressed in Chapters 5 and 6.

### 4.2.1 Defining Acceptance Criteria

Acceptance criteria provide necessary information for evaluating equipment integrity to help ensure that corrective actions are taken when necessary. Acceptance criteria also define the limits for a specific ITPM activity.

The criteria can be expressed qualitatively or quantitatively. An example of a qualitative criterion is "no missing or bent pipe supports." An example of a

# 4. INSPECTION, TESTING, AND PREVENTIVE MAINTENANCE 47

quantitative acceptance criterion is "the wall loss is less than the stated corrosion allowance of 0.125 in." Chapter 8 provides additional information on defining and using acceptance criteria.

### 4.2.2 Equipment and ITPM Task Results Documentation

Other necessary components of the ITPM program are proper documentation and systems to manage this documentation. In general, two types of documentation are needed: (1) equipment documentation and (2) ITPM task results documentation.

*Equipment Documentation.* Some of the same equipment information used for task selection is needed for ITPM task execution. For example, personnel executing ITPM tasks should review relevant information in the equipment file. Information of interest includes (1) ITPM history, (2) equipment details (e.g., specific components requiring inspection), (3) ITPM task details (e.g., TMLs, inspection technique), and (4) acceptance criteria. Reviewing equipment-specific information helps ensure that (1) any questionable areas are thoroughly examined during the task, (2) specifics related to the inspection are known (e.g., inspection technique, sampling criteria), and (3) the inspection is performed consistently (e.g., taking thickness readings at the same location).

*ITPM Task Results Documentation.* The second type of documentation, ITPM task results, is generated as part of executing the task. For most ITPM tasks, the results are documented each time the task is executed, regardless of the organization/department performing the task. Documenting the task results provides data needed to (1) determine the integrity of the equipment and identify equipment deficiencies, (2) perform evaluations needed to adjust task frequency (e.g., remaining life calculations), and (3) spot trends that might help predict failures.

In addition, documenting ITPM tasks provides regulators and auditors evidence that the tasks have been performed as scheduled by qualified personnel. While the content of the results documentation will vary somewhat based on the ITPM task, some minimum documentation requirements (Reference 4-1) are:

- Equipment identifier, such as the asset number, serial number, or equipment tag number
- Date the ITPM task was performed
- Name(s) of the person(s) performing the ITPM task
- Description of the ITPM task performed (e.g., magnetic particle testing, ultrasonic testing measurements, radiography, replacement of impeller)
- Results of the ITPM task

For a more effective MI program, facilities should document more information. The additional information documented usually depends on the type of task (i.e., inspection, testing, or PM) and the type of equipment. Some examples of additional information that could be included in the ITPM results documentation are:

- The equipment's fitness-for-service evaluation
- Any materials or spare parts used
- Any QA records associated with the task (e.g., verification of replacement part materials of construction)
- The qualification records of personnel performing the task (e.g., API 510 inspector qualification record)
- Any detailed data associated with the task (e.g., ultrasonic thickness [UT] readings, vibration spectrums, photographs of tank/vessel internals), especially any data that should be referenced the next time that the task is performed
- Any identified equipment deficiencies
- Any recommended corrective actions
- Any remaining life calculations (or similar calculations or assessment)
- The next scheduled task date

However, an exception to the ITPM task results documentation practices may be taken for certain predesignated ITPM tasks. Often, for these tasks, the results can be better managed by documenting negative results only (i.e., document by exception). Documentation by exception should be used cautiously, and the use of this documentation approach should be clearly stated in the ITPM plan (Reference 4-7). This documentation approach is best used:

- For tasks with short intervals (e.g., daily, weekly)
- When trending of data is not needed to determine equipment integrity
- When data history is not needed to determine inspection requirements
- When appropriate control or audit mechanisms are in place to ensure that these tasks are performed as intended

The methods and formats used to document task results are varied. The results can be documented electronically or on paper. Many times, paper documentation is used when (1) entering the information into a computer system is not cost-effective and/or (2) the available computer system is not capable of managing the results. For example, facilities lacking specialized inspection data management software often find it cost prohibitive to manually enter thickness measurement data into the CMMS. Fortunately, some inspection software packages (i.e., programs that can better manage data) are becoming more economical and more common. Additional information on use of CMMS and other software packages is provided in Chapter 12.

Facilities also need to decide on the most applicable format for collecting and documenting the task results. In general, format is influenced by the type of task and type of information to be recorded. For example, a lubrication route is usually best documented using a check sheet format. On the other hand, internal inspections of pressure vessels are usually documented in formal inspection

## 4. INSPECTION, TESTING, AND PREVENTIVE MAINTENANCE 49

reports that contain all data collected (e.g., wall thickness measurements). The method and format chosen for each ITPM task should be communicated to involved personnel. Chapter 6 includes more details on procedure formats.

Facilities should develop a policy that defines the retention requirements for ITPM results documentation. This policy should ensure that ITPM results for equipment covered by process safety regulations are retained for the life of the process, especially ITPM results needed to show that equipment has been maintained in accordance with RAGAGEPs. However, retaining results for all ITPM tasks, especially frequently performed tasks (e.g., daily or weekly tasks), is not practical or necessary. When developing a policy for retaining and discarding documentation, facility personnel should ensure that:

- Documents needed to demonstrate that equipment is maintained, inspected, tested, and operated in a safe manner (e.g., complies with applicable RAGAGEPs) are retained.
- A sufficient number of documents are retained to demonstrate that the ITPM tasks are being performed as required and the results are being appropriately documented.
- Discarded documents are not needed to execute future ITPM tasks or other maintenance activities (e.g., vessel repair).
- Legal, regulatory, company, or other requirements that specify record retention requirements are met.

### 4.2.3 ITPM Task Implementation and Execution

Once the ITPM plan and schedule are developed, the ITPM tasks have to be implemented and then executed as required by the ITPM plan and schedule. To implement ITPM tasks, facilities need to ensure that (1) procedures are developed, (2) personnel are trained, and (3) sufficient resources are provided for ITPM program startup and ongoing execution of the ITPM tasks. Chapters 5 and 6 provide information on training and procedures. In addition, Chapter 10 discusses resource issues.

Once the tasks are being performed, facilities need to ensure that the task results and task schedule are appropriately managed. A key part of task management is ensuring that results are reviewed for equipment deficiencies. The following subsection of this chapter and Chapter 8 provide additional information on reviewing tasks and managing equipment deficiencies. In addition, the task scheduling process needs to (1) track whether tasks are executed when required and that any deferred tasks are properly managed and (2) include activities for reviewing and optimizing ITPM tasks and their frequencies. Subsection 4.2.5 addresses task scheduling issues.

### 4.2.4 ITPM Task Results Management

Unless facilities are prepared to manage them, ITPM task results can easily become overwhelming. Establishing a system for managing the task results can provide tremendous benefits, such as:

- Improved equipment integrity, resulting in improved process safety, environmental compliance, and equipment reliability
- Increased employee buy-in of the MI program (employees like to see that their work efforts are valued)
- Improved regulatory compliance (unresolved ITPM task deficiencies are typical items identified during MI program audits)

If task results are not managed, facilities are likely to miss opportunities to (1) identify and properly manage equipment deficiencies and (2) optimize the ITPM schedule. Therefore, successful MI programs typically include a management system that assigns personnel the responsibility to (1) review task results, (2) perform any needed calculations/evaluations, (3) update the ITPM schedule, and (4) initiate corrective actions.

To help ensure proper review of task results, personnel with appropriate skills and knowledge should be assigned to this responsibility. Specifically, personnel reviewing ITPM task results will need to:

- Verify that the ITPM activity was completed (as required) and perform a preliminary review of the data and results. Specifically, data should be reviewed for anomalies so that suspicious information can be verified or corrected.
- Evaluate and analyze the results and recommendations. This may include entering data into software for analysis, performing (or initiating the performance of) evaluations (e.g., fitness for service [FFS]) or calculations (e.g., corrosion rates, next inspection date), and/or reviewing recommendations for practicality and effectiveness.
- Identify (or initiate a review to determine) equipment deficiencies. For some ITPM activities, this step may simply involve comparing the task results to established acceptance criteria (e.g., minimum thickness values, no leaks from seals). For other ITPM activities, identifying deficiencies may require personnel (or a team) with specific expertise (e.g., pressure vessel expertise). In addition, recommendations for resolving any deficiencies may be developed. Once a deficiency is identified, the facility's equipment deficiency process is used to manage the deficiency. Chapter 8 discusses the equipment deficiency process in detail.
- Track equipment deficiencies. Usually, a list or log of equipment deficiencies is maintained to help ensure that they are resolved. Again, Chapter 8 provides additional information on this topic.
- Resolve minor issues. Sometimes, ITPM activities identify equipment damage/degradation issues that are not (or have not yet resulted in a level of damage/degradation to be considered) equipment deficiencies (e.g., minor rust, slightly higher than normal vibration reading). The ITPM task result review should ensure that necessary corrective actions are implemented. When no corrective actions are required, the review process should ensure that this decision and its basis are documented.

## 4. INSPECTION, TESTING, AND PREVENTIVE MAINTENANCE

A prompt review of task results (during outage windows) helps ensure that any corrective actions that require the equipment to be out of service can be implemented before equipment is returned to service. In addition, prompt review of reports from frequently performed tasks helps ensure that problems noted by employees are addressed in a timely manner. Because of the large number of ITPM reports that can be generated, personnel should be trained to highlight out-of-the-ordinary results. In some facilities, reports of out-of-the-ordinary results are routed differently from routine reports. Similarly, firms contracted to perform ITPM activities can be instructed to highlight areas of concern in a summary report or cover letter that can help keep this important information from being lost or overlooked.

### 4.2.5 Task Schedule Management

Ideally, ITPM tasks should be performed as scheduled; however, tasks might occasionally be delayed or missed. Facilities must strive to minimize this practice. To help ensure that delayed tasks are properly managed, a management system can be set up to verify that:

- Task delays are warranted
- The risks associated with delays are understood
- Appropriate management of change (MOC) reviews are performed
- Task delays are approved by an appropriate level of management
- Interim measures needed to ensure successful continued operation of the equipment (e.g., increasing visual inspections of a tank when a formal inspection is delayed) are identified and implemented

Another system that facilities can use to help manage the ITPM task schedule is to define acceptable grace periods for ITPM tasks. These grace periods define the acceptable time frame for performing the task. A grace period is intended to assist facilities in managing the ITPM task schedule to align with available resources (e.g., maintenance labor) and operations. The actual grace period typically varies with the task interval (e.g., 1 day for weekly tasks, 1 week for quarterly tasks).

Because RAGAGEPs rarely define an acceptable grace period, organizations should be cautious when establishing and using grace periods. Otherwise, regulators and auditors may get the impression that the grace periods are routinely being used to extend or avoid performing ITPM tasks. For example, if monthly tasks have a 1-week grace period, and tasks are always performed every 5 weeks, the task may wind up being performed 10 times in a year rather than the intended 12 times. One technique to help to prevent this from occurring is to maintain task due dates on defined baseline due dates. For example, monthly tasks can be set with due dates based on a monthly calendar (e.g., the first week of each month) rather than 1 month after the task was last performed.

Another feature of task schedule management is the adjustment of task frequencies based on ITPM results. For example, the RAGAGEPs for vessels, tanks, and piping include formulas for calculating the remaining life of the equipment (based on the inspection results) and criteria for adjusting task frequencies (References 4-5, 4-8, 4-9, and 4-10). Similarly, ITPM task results for other types of equipment can be used to adjust task frequencies. For example, testing frequencies can be increased or decreased based on the results of function tests. Protocols should be established for frequency adjustment. Any deviations from the protocol should be approved through the facility's MOC program (or a similar change management process).

One common problem encountered with task frequency adjustment is the lack of baseline readings. Facilities do not always obtain baseline readings when equipment is initially commissioned. Therefore, some nominal value (e.g., nominal thickness of the steel plate used to manufacture a pressure vessel wall) is assigned and used to calculate initial deterioration rates (e.g., corrosion rates), and ultimately to determine the remaining life. Problems can arise when the true value deviates from the assigned nominal value, leading to overly pessimistic or optimistic calculations of the remaining life. For ITPM tasks with low frequency (i.e., a long interval between tasks, such as 5 or 10 years), obtaining accurate baseline readings improves a facility's confidence in the remaining life calculations. In addition, providing baseline readings offers some QA of new installations.

Ensuring that task results are used to adjust task frequencies is important because it provides an opportunity to (1) increase the task frequency to help ensure that failure does not occur before the next execution of the task, (2) decrease the task frequency to help reduce the MI program resource requirements, and (3) eliminate tasks that are unnecessary (e.g., eliminate thickness measurements on equipment in processes that are not corrosive to the material of construction). Most organizations benefit from having a procedure that both defines the responsibility for reviewing results/optimizing task frequencies and provides criteria for when frequencies can be adjusted.

### 4.2.6 ITPM Program Monitoring

Monitoring ITPM program performance includes establishing appropriate performance measures, such as percent of ITPM tasks performed on time, number of equipment items with ITPM tasks, or number of equipment items with outstanding equipment deficiencies. Chapter 12 provides additional information on performance measures for MI programs, including suggested performance measures related to the ITPM program.

Periodic reports regarding the ITPM program's performance should be generated and reviewed by management. One report to receive management attention is an overdue ITPM task report; management must be kept aware when ITPM tasks are not completed as scheduled. Management then should determine the causes of the late/deferred ITPM tasks and ensure that necessary corrective actions are developed and implemented in a timely manner.

## 4.3 ITPM PROGRAM ROLES AND RESPONSIBILITIES

While the assignment of roles and responsibilities will vary between different organizations, basic roles and responsibilities should be assigned to appropriate positions in the organization. Many of the roles and responsibilities for the ITPM program are assigned to the inspection and maintenance departments. However, involvement of production personnel is needed and advisable, as described in Section 4.1. In addition, project and maintenance storeroom personnel may need to be involved.

Example roles and responsibilities for the ITPM program are provided in Tables 4-5 and 4-6. Table 4-5 indicates example roles and responsibilities for the ITPM task planning phase and Table 4-6 indicates example roles and responsibilities for the ITPM task execution and monitoring phase. The matrixes designates assignments to personnel with "R" as the person(s) responsible for the activity, "A" as the approver of the work or decisions made by the responsible party, "S" as persons supporting the responsible party in completing the activity, and "I" as persons informed when the activity is completed or delayed.

## 4.4 REFERENCES

4-1. Occupational Safety and Health Administration, *Process Safety Management of Highly Hazardous Chemicals*, 29 CFR Part 1910, Section 119, Washington, DC, 1992.

4-2. ABSG Consulting Inc., *Mechanical Integrity, Course 111*, Process Safety Institute, Houston, TX, 2004.

4-3. Occupational Safety and Health Administration, *Pressure Vessel Guidelines*, Technical Manual TED 1-0.15A, Section IV, Chapter 3, Washington, DC, 1999.

4-4. Wulpi, D., *Understanding How Components Fail*, 2nd Edition, ASM International, Materials Park, OH, 1999.

4-5. American Petroleum Institute, *Piping Inspection Code: Inspection, Repair, Alteration, and Rerating of In-Service Piping*, API 570, Washington, DC, 2003.

4-6. American Petroleum Institute, *Inspection Practices for Piping System Components*, API RP 574, Washington, DC, 1998.

4-7. Occupational Safety and Health Administration, *Clarification on the Documentation of Inspections and Tests Required Under the Mechanical Integrity Provisions*, Correspondence letter to Mr. Sylvester W. Fretwell, Lever Brothers Company, dated September 16, 1996.

4-8. American Petroleum Institute, *Pressure Vessel Inspection Code: Maintenance Inspection, Rating, Repair and Alteration*, API 510, Washington, DC, 2003.

4-9. American Petroleum Institute, *Tank Inspection, Repair, Alteration, and Reconstruction*, API 653, Washington, DC, 2003.

4-10. National Board of Boiler and Pressure Vessel Inspectors, *National Board Inspection Code*, 2004 Edition, Columbus, OH, 2005.

4-11. ABSG Consulting Inc., *Overview of Condition Monitoring Techniques for ABS Technology*, Houston, TX, 2004.

## Additional Source

American Petroleum Institute, *Damage Mechanisms Affecting Fixed Equipment in the Refining Industry*, API RP 571, Washington, DC, 2003.

# 4. INSPECTION, TESTING, AND PREVENTIVE MAINTENANCE

**TABLE 4-5**
**Example Roles and Responsibilities Matrix for the ITPM Task Planning Phase**

| Activity | Inspection and Maintenance Department Personnel | | | | | | | Production Department Personnel | | | | | Other Personnel | | | | | |
|---|---|---|---|---|---|---|---|---|---|---|---|---|---|---|---|---|---|---|
| | Inspection Manager | Maintenance Manager | Maintenance Engineers | Maintenance Supervisors | Inspectors | Maintenance Technicians | Maintenance Planner/Scheduler | Production Manager | Area Superintendent/Unit Manager | Production/Process Engineers | Production Supervisors | Operators | Plant Manager | Project Managers/Engineers | Storeroom Personnel | Process Safety Coordinator | Manufacturers and Outside SMEs | Outside Contractors |
| **Task Selection/Frequency Determination** | | | | | | | | | | | | | | | | | | |
| • Inspection tasks | R/A | | S | S | S | | S | S | R | S | S | | — | | | — | S | S |
| • Testing tasks | | R/A | S | S | | S | S | S | S | S | S | | — | | | — | S | S |
| • Preventive maintenance tasks | | R/A | S | S | R | S | | S | S | S | S | | — | | | — | S | S |
| • Sampling Criteria | A | | S | | R | | | | | | | | | | | — | | |
| **Task Assignments (at Department or Craft Level)** | | | | | | | | | | | | | | | | | | |
| • Inspection tasks | R/A | | | | S | | S | S | | | | | — | | | — | | |
| • Testing tasks | | A | | R | | S | S | | | | | | | | | = | | |
| • Preventive maintenance tasks | | A | | R | | S | S | | | | | | | | | — | | |
| • Operations ITPM tasks | | | | | | | | | R | | | S | — | | | — | | |
| • Task Scheduling | A | A | | S | S | | R | S | S | | | | — | | | — | | |

## TABLE 4-6
## Example Roles and Responsibilities Matrix for the ITPM Task Execution and Monitoring Phase

| Activity | Inspection and Maintenance Departments | | | | | | | Production Department Personnel | | | | | Other Personnel | | | | | |
|---|---|---|---|---|---|---|---|---|---|---|---|---|---|---|---|---|---|---|
| | Inspection Manager | Maintenance Manager | Maintenance Engineers | Maintenance Supervisors | Inspectors | Maintenance Technicians | Maintenance Planner/Scheduler | Production Manager | Area Superintendent/Unit Manager | Production/Process Engineers | Production Supervisors | Operators | Plant Manager | Project Managers/Engineers | Storeroom Personnel | Process Safety Coordinator | Manufacturers and Outside SMEs | Outside Contractors |
| **Acceptance Criteria Determination** | | | | | | | | | | | | | | | | | | |
| • Inspection tasks | A | | S | | R | | | | | S | | | | | | | S | S |
| • Testing tasks | | A | R | S | | S | | | | S | | | | | | | S | S |
| • Preventive maintenance tasks | | A | R | S | | S | | | | S | | | | | | | S | |
| **Task Execution** | | | | | | | | | | | | | | | | | | |
| • Inspection tasks | A | | | | R | | | | | | | | – | | | – | R | R |
| • Testing tasks | | A | | | | R | S | | | | S | R | – | | | – | R | R |
| • Preventive maintenance tasks | | | | | | R | S | | | | S | – | – | | | – | R | |
| • Operations ITPM tasks | | | | | | | | | | | P | R | | | | | | |
| **Equipment Documentation Generation and Maintenance** | | | | | | | | | | | | | | | | | | |
| • Design and construction information | • | | R | | | | | | | S | | | | R/A | | | | |
| • Service history | R/A | | | | S | S | | | | R/A | S | S | | | | | | |

## 4. INSPECTION, TESTING, AND PREVENTIVE MAINTENANCE

**TABLE 4-6 (Continued)**

| Activity | Inspection and Maintenance Departments | | | | | | | Production Department Personnel | | | | | Other Personnel | | | | | |
|---|---|---|---|---|---|---|---|---|---|---|---|---|---|---|---|---|---|---|
| | Inspection Manager | Maintenance Manager | Maintenance Engineers | Maintenance Supervisors | Inspectors | Maintenance Technicians | Maintenance Planner/Scheduler | Production Manager | Area Superintendent/Unit Manager | Production/Process Engineers | Production Supervisors | Operators | Plant Manager | Project Managers/Engineers | Storeroom Personnel | Process Safety Coordinator | Manufacturers and Outside SMEs | Outside Contractors |
| ♦ ITPM history | A | A | | | R | R | S | | | | | | | | – | | | |
| ♦ Maintenance history | A | A | | | R | R | S | | | | | | | | – | | | |
| ♦ Vendor-supplied information | | | R | | | | | | | | | | | | – | | | |
| ITPM Task Results Documentation Management | | | | | | | | | | | | | | | | | | |
| ♦ Documentation requirements | R | R | S | S | S | | | | R | | R | | | R/A | | | | |
| ♦ Results documentation | | | S | R | R | S | S | | | | | | | | | | | |
| ITPM Task Results Management | – | – | R | R | R | S | S | | S | | R | | | | | | | |
| ITPM Schedule Management | | | | | | | | | | | | | | | | | | |
| ♦ Delayed or omitted tasks | R/A | R/A | S | R | S | S | S | A | S | | S | S | A | | | | | |
| ♦ Task frequency adjustment | A | A | S | R | R | S | S | A | S | | S | S | – | | – | – | R | R |
| ITPM Program Monitoring | R/A | R/A | | S | S | S | S | – | – | | – | – | – | – | – | – | R | R |

**Appendix 4A. Common Predictive Maintenance and Nondestructive Testing Techniques**

Detecting equipment failures before they occur is a primary objective of the MI program. Because equipment failures are many times preceded by an advanced warning period, QA and ITPM programs often include predictive maintenance and NDT techniques aimed at detecting and recognizing these warnings. Both predictive maintenance and NDT are considered to be condition monitoring (CM) activities. Below are brief descriptions of some general CM categories and an overview of different types of CM tasks (Reference 4-11). The CD accompanying this book contains examples of the CM techniques listed below.

*Temperature Measurement.* Temperature measurement (e.g., temperature indicating paint, thermography) helps detect potential failures related to a temperature change in equipment. Measured temperature changes can indicate problems such as excessive mechanical friction (e.g., faulty bearings, inadequate lubrication), degraded heat transfer (e.g., fouling in a heat exchanger), and poor electrical connections (e.g., loose, corroded, or oxidized connections).

*Dynamic Monitoring.* Dynamic monitoring (e.g., spectrum analysis, shock pulse analysis) involves measuring and analyzing energy emitted from mechanical equipment in the form of waves, such as vibrations, pulses, and acoustic effects. Measured changes in the vibration characteristics from equipment can indicate problems such as wear, imbalance, misalignment, and damage.

*Oil Analysis.* Oil analysis (e.g., ferrography, particle counter testing) can be performed on different types of oils, such as lubrication, hydraulic, or insulation oils. Oil analysis can indicate problems such as machine degradation (e.g., wear), oil contamination, improper oil consistency (e.g., incorrect or improper amount of additives), and oil deterioration.

*Corrosion Monitoring.* Corrosion monitoring (e.g., coupon testing, electrical resistance testing, linear polarization, electrochemical noise measurement) helps provide an indication of the extent of corrosion, the corrosion rate, and the corrosion state (i.e., active or passive corrosion state) of a material. The CD accompanying this book contains examples of corrosion CM techniques.

*Nondestructive Testing.* NDT involves performing tests that are noninvasive to the test subject (e.g., x-ray, ultrasonic). Many of these tests can be performed while the equipment is on line.

*Electrical Testing and Monitoring.* Electrical CM techniques (e.g., high potential testing, power signature analysis) involve measuring changes in system properties such as resistance, conductivity, dielectric strength, and potential. These techniques are helpful in detecting such conditions as electrical insulation deterioration, broken motor rotor bars, and shorted motor stator lamination.

## 4. INSPECTION, TESTING, AND PREVENTIVE MAINTENANCE

*Observation and Surveillance.* Observation and surveillance CM techniques (e.g., visual, audio, and touch inspections) are based on human sensory capabilities. Observation and surveillance can supplement other CM techniques in detecting such problems as loose/worn parts, leaking equipment, poor electrical/pipe connections, steam leaks, pressure relief valve (PRV) leaks, and surface roughness changes.

*Performance Monitoring.* Monitoring equipment performance is a form of CM that predicts problems by monitoring changes in variables such as pressure, temperature, flow rate, electrical power consumption, and/or equipment capacity.

# 5
# MI TRAINING PROGRAM

An important ingredient of an effective mechanical integrity (MI) program is personnel training. Proper training helps to ensure that only qualified personnel perform MI tasks and that MI tasks are performed appropriately and consistently (i.e., with fewer opportunities for human errors). Reducing human errors can greatly reduce the overall rate of equipment failures. Figure 5-1 provides a flow chart for the development and implementation of a training program.

This chapter addresses training and qualification of the facility's MI workforce as well as the selection of qualified contractor personnel. The following topics are included:

- Skills/knowledge assessment
- Training for new and current workers
- Verification and documentation of training effectiveness
- Certification, where applicable
- Ongoing and refresher training
- Training for technical personnel
- Contractor issues
- Roles and responsibilities

Equipment-specific training and/or certification requirements are discussed in more detail in Chapter 9 and in the equipment-specific matrices on the CD accompanying this book. Further information on training is available in ANSI Z490.1, *Criteria for Accepted Practices in Safety, Health and Environmental Training*.

## 5.1  SKILLS/KNOWLEDGE ASSESSMENT

Formal skills/knowledge assessments can be used to identify gaps in knowledge of the current workforce and to evaluate new hires. Knowing where weaknesses exist allows a facility to develop training or seek outside assistance to address the greatest training and qualification needs. The following describes a

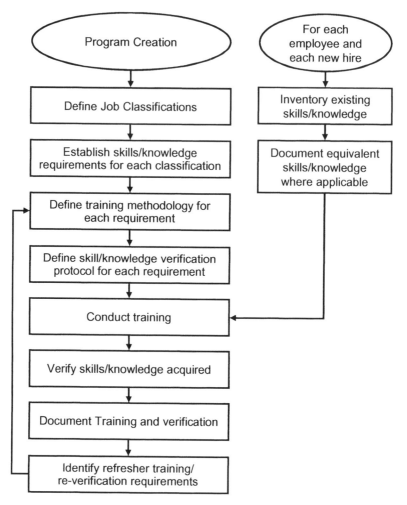

FIGURE 5-1  Training flow chart.

skills/knowledge assessment technique that has been successfully applied at several facilities:

1. Review the MI program components, such as inspection, testing, and preventive maintenance (ITPM) plans, quality assurance (QA) plans, and common repair activities, to identify and list MI tasks to be performed.
2. Develop a list of the job classifications for the personnel who will perform the tasks. Note that some tasks will be performed by personnel outside the traditional maintenance workforce (e.g., storeroom workers, buyers, process operators), and other tasks may be designated for contractors. When possible, identify job classifications and subclassifications that are independent of pay classifications. For example, a facility may have one

## 5. MI TRAINING PROGRAM 63

pay grade for the instrument and electrical workforce; however, for training purposes, classifications (e.g., mechanical maintenance technician, I&E technician) and subclassifications (e.g., millwright, DCS technician) may be different or more specific.
3. Solicit input from experienced, knowledgeable personnel when assembling lists of the skills/knowledge required within each job classification and subclassification. Certification requirements (particularly for welders and fixed equipment inspectors) and other skills mandated by recognized and generally accepted good engineering practice(s) (RAGAGEPs) and/or jurisdictional requirements may also be included. Add these skills/knowledge items to the list of MI tasks identified in Step 1.
4. To maintain a comprehensive skills/knowledge list, add any additional training that is required for other process safety programs, process overviews and hazards, safe work practices, and any other required training (e.g., management of change [MOC], emergency response, hazardous material [HAZMAT] training).
5. Survey the workforce to determine which skills/knowledge areas need improvement. Consider using a survey format that asks personnel to evaluate both their personal capabilities and those of their co-workers. Note: Ensure that any surveys of co-worker abilities are done in a nonpersonal manner. See Appendix 5A for a sample survey format.
6. Survey supervisory personnel to obtain their input on the skill/knowledge base of the workforce.
7. Also, consider what, if any, skills/knowledge the next worker hired will need to develop. If the maintenance workforce has experienced craft backgrounds, the needed training will differ from a facility whose maintenance workforce is primarily filled by less experienced personnel or by process operators transferring into the department.
8. Identify skills/knowledge gaps for the existing workforce. Also, identify requirements for promoted personnel and new hires.
9. Determine whether or not to continue performing relatively infrequent tasks with the in-house workforce.

Often, a facility's existing workforce has been performing MI tasks for many years. Therefore, personnel may show some resistance to the concept of skills/knowledge assessment. Some of the following suggestions should help overcome any resistance:

- Present skills/knowledge assessment in a positive light. Most workers will welcome the chance to upgrade their skills/knowledge.
- Involve experienced workers when identifying necessary skills/knowledge requirements for each job classification.
- Evaluate general gaps in the workforce skills/knowledge in addition to a worker's personal needs.

- Present the outline for the entire training program. Demonstrate that skills/knowledge assessment is just one of the steps that management is taking to enhance the training program.
- Don't use the skills/knowledge assessment (or other aspects of the training) as a tool to discipline or fire ineffective workers.

## 5.2 TRAINING FOR NEW AND CURRENT WORKERS

For each of the improvement needs identified, determine whether skill enhancement (or skill remediation) can best be accomplished by enhancing procedures, training, or both. Chapter 6 has more information on MI procedures.

Several training approaches are available to most facilities. Training can occur in a variety of locales: in classrooms, in front of a computer, on shop floors, in the field, at community colleges, at a vendor school, and/or through a correspondence school. Facilities should (1) determine which training approaches would be most effective for meeting the MI objectives and (2) ensure that resources are allocated to support the training. Often, facilities use a combination of training approaches. For example, safe work practices may be taught in a classroom, whereas training on compressor maintenance may require field work in addition to, or instead of, classroom education. Table 5-1 compares some of the typical strengths and potential weaknesses inherent in some common training approaches.

Training can be provided by a variety of instructors: experienced personnel, engineers, vendors, and contractors. Selection and training of the trainers are very important. Facilities should ensure that trainers know and follow accepted work procedures. Trainers should be selected on the basis of their knowledge of the training topic and on their ability to effectively transfer knowledge to others. A highly skilled mechanic does not necessarily have the talent and ability to be a good teacher. Many facilities implement a training program for the trainers. Some facilities send designated trainers to manufacturers' training classes and rely on those trainers to, in turn, pass that knowledge on to the rest of the workforce.

Many facilities create standard guides to be used to facilitate each requirement of a training program. The CD accompanying this book includes a sample training guide for repairing a mechanically sealed pump.

## 5.3 VERIFICATION AND DOCUMENTATION OF TRAINING EFFECTIVENESS

Facilities should consider establishing criteria for judging whether training has enhanced a person's skills/knowledge. A variety of means can be used for training verification: some skills can be verified with a written exam; however, field/shop demonstration is often more appropriate (and usually more popular with the trainees). Field training, in particular, should have specific objective criteria for judging proficiency.

# 5. MI TRAINING PROGRAM

**TABLE 5-1**
**Training Approach Considerations**

| Approach | Typical Strengths | Potential Weaknesses |
|---|---|---|
| Self-paced training (e.g., computer-based training packages) | • Consistent information delivered<br>• Verification that training is understood for each individual<br>• Training schedules can be catered to individuals<br>• May minimize "trainer" expenses (in the long term) | • Modules may not be relevant to actual tasks and may not be consistent with company practices (may require review and customization of modules to ensure relevance)<br>• No interaction with a live instructor or classmates<br>• No field skills verification<br>• Potentially high startup expenses |
| Classroom sessions | • Consistent information delivered<br>• Curriculum will be relevant and can be adjusted to the needs of the workforce<br>• Interaction with classroom instructor and with classmates<br>• Acquired knowledge can be tested readily (by trainer/student interaction as well as by written tests) | • Training material development expenses<br>• Trainer and student schedules must be coordinated<br>• Requires a system to ensure that employees who miss training sessions are trained in make-up sessions<br>• Sessions may emphasize group needs, but will not accommodate individual needs<br>• No field skills verification |
| On-the-job training | • Training can be readily adjusted to meet individual needs<br>• Curriculum will be relevant to day-to-day activities<br>• Experienced (knowledgeable) resource person is readily available to coach and answer questions<br>• Field skills can be verified by the trainer | • Inconsistent information -- a trainee may work with multiple people and become confused if different personnel approach tasks differently (consider designating a limited number of personnel to be trainers)<br>• Incomplete information -- unstructured day-by-day training may never encounter some tasks (consider establishing checklists and other tools to ensure that all training topics are covered consistently for each employee)<br>• Improper practices and unconventional "wisdom" may be carried from employee to employee, leading to a workforce-wide deficiency (consider supplementing on-the-job training with other methods to address workforce-wide deficiencies)<br>• Training verification may appear subjective (establish specific criteria for skills verification)<br>• Temporary loss or less productive use of skilled workers involved in training the new employees |

The training program should include provisions for documenting the dates of the training, how the skills/knowledge were verified (e.g., test, observation), and the results of the verification activity (e.g., test score, pass/fail). Training requirements and training completed should be documented for each individual. Many facilities establish a database that lists each employee and the employee's job classification(s) in a matrix versus the skills/knowledge and/or training classes required. Table 5-2 shows a format (and part of a matrix) for a general electrician. On the CD accompanying this book are samples of skills/knowledge lists for a general electrician and for a general mechanic. Backup documentation (e.g., completed written tests, completed checklists) should be retained for each entry in the employee skills/knowledge matrix to demonstrate that the employee has met the established criteria. Dates of training should be logged and, as appropriate, systems should be established to keep personnel training current (especially for training that requires periodic refreshing). Some facilities incorporate training reminders into the work order system or into another computer program that flags "training required" dates. Numerous training software databases are commercially available that incorporate many or all of these features.

In some cases, employees are brought into a facility with existing skills. To ensure that the training database is not compromised by inconsistencies in documentation, facilities should establish criteria for "equivalent" ability/knowledge demonstration. For instance, an electrician having a journeyman status who is hired into the facility may be exempt from some of the general electrical training that would be required of a less-skilled new hire. Similarly, a facility operator who transfers to the Maintenance department may have already received facility-specific or process-specific training that would be required of a new hire from outside of the facility.

## 5.4 CERTIFICATION

Although many MI skills have no certification requirements, facilities need to be aware of, and make use of, widely recognized certification requirements when they do apply. In particular, certifications are available for high-profile MI activities such as fixed equipment inspection and welding. In some cases, jurisdictional rules (e.g., state laws) also include certification requirements. Although codes and laws are subject to change, the trend is to require more certification for maintenance and engineering work. In addition, for some codes (e.g., welding), systems need to be established to track performance in order to keep certifications current. Table 5-3 contains a list of some of the widely accepted certifications available for MI work. For those skills that have no external certification requirements, many companies establish qualification or internal certification requirements as part of their training verification and documentation process.

# 5. MI TRAINING PROGRAM

## TABLE 5-2
## General Electrician Training Matrix

| | Requirement | | | | | | | | |
|---|---|---|---|---|---|---|---|---|---|
| | Introduction to Basic Electricity | Fundamentals of DC Circuits | Fundamentals of AC Circuits | Safety in Electrical Maintenance | Electrical Drawings | Electrical Test Equipment | DC Troubleshooting Techniques | AC Troubleshooting Techniques | Electronic Process Transmitters |
| Harris, George | 1/15/00 | 3/17/00 | 4/21/00 | 8/31/99 | 11/11/99 | 9/15/99 | 1/19/01 | 10/31/02 | 5/19/03 |
| Jones, Mick | 12/20/99 | 3/17/00 | 8/15/98 | 8/31/99 | 11/11/99 | 9/15/99 | 6/13/01 | 10/31/02 | 5/19/03 |
| Leonard, John | Waived (based on experience) | Waived (based on experience) | Waived (based on experience) | | Waived (based on experience) | Waived (based on experience) | Waived (based on experience) | Waived (based on experience) | 5/19/03 |
| Olsen, Paul | 12/20/99 | 3/17/00 | 4/21/00 | 8/31/99 | 11/11/99 | 9/15/99 | 6/13/01 | 10/31/02 | 5/19/03 |
| Richards, Sam | 1/15/00 | 3/17/00 | 5/1/00 | 8/31/99 | 11/11/99 | 9/15/99 | 1/19/01 | 10/31/02 | 5/19/03 |
| Starkey, Greg | 1/15/00 | | 4/21/00 | 8/31/99 | 11/11/99 | 9/15/99 | 6/13/01 | 10/31/02 | 5/19/03 |
| Watts, Bob | 12/20/99 | 3/17/00 | 8/15/98 | 8/31/99 | 11/11/99 | 9/15/99 | 6/13/01 | 10/31/02 | 5/19/03 |
| Wyman, George | Waived (based on experience) | | 5/1/00 | 2/12/00 | 11/11/99 | 9/15/99 | 6/13/01 | 10/31/02 | 5/19/03 |

**TABLE 5-3**
**Widely Accepted MI Certifications**

| Skill | Organization | Certificate or Standard Used |
|---|---|---|
| Atmospheric storage tank inspection | API | API 653 |
| Boiler, pressure vessel, and piping inspection | National Board of Boiler and Pressure Vessel Inspectors (NBBPVI) | National Board (NB)-23 |
| Corrosion prevention and control | NACE International | • Cathodic Protection Specialist<br>• Corrosion Specialist<br>• Chemical Treatment Specialist<br>• Materials Selection/Design Specialist<br>• Protective Coatings Specialist |
| Inspection supervision, technique selection, procedure preparation and approval, code interpretation | American Society of Non-destructive Testing (ASNT) | ASNT Central Certification Program (ACCP) Level III |
| NDT techniques | ASNT | SNT-TC-1-A, for various individual types of inspections and tests |
| NDT techniques, equipment calibration, interpretation of results | ASNT | • ACCP Level II<br>• Magnetic particle testing<br>• Liquid penetrant testing<br>• Radiographic testing<br>• Ultrasonic testing<br>• Visual testing |
| Piping inspection | API | API 570 |
| Pressure vessel inspection | API | API 510 |
| Welding | ASME | Boiler and Pressure Vessel Code (BPVC) Section IX, Welding (for qualifications specific to metals, rods, and weld positions) |
| | American Welding Society (AWS) | Certified Welder |
| Weld inspector | AWS | • Certified Welding Inspector<br>• Senior Certified Welding Inspector |

## 5.5 ONGOING AND REFRESHER TRAINING

Updating the task and skill lists developed for the skills/knowledge assessment should be completed as new equipment and/or new maintenance techniques are introduced to the facility. Ideally, these updates should occur as part of the facility's MOC program. Traditionally, many facilities have provided training on new equipment with introductory classes taught by vendors and/or project engineers associated with the equipment. In other cases, companies have sent employees to factory schools. Such courses may still be appropriate; however, the training program should also consider how the next person will acquire these new skills. That is, how will the facility train personnel who did not attend the class? The following questions can be used when evaluating these introductory classes:

- What knowledge and/or skills did class attendees acquire?
- Are sufficient resources (e.g., knowledge, reference materials) available to provide this training in-house?
- Was the class useful and should it be repeated in the future?
- Does it make sense to train each worker individually (e.g., one-on-one with an experienced worker or perhaps at a vendor school) or should several workers be trained together?

In addition, facilities should provide additional training if (1) employee performance indicates that more training is necessary, (2) employees request additional training, (3) procedures/job tasks change, or (4) a process and/or its hazards change.

Furthermore, for some topics, refresher training makes sense (e.g., based on employee needs/desires) and, in some cases, is required (e.g., by jurisdictional laws, by RAGAGEPs). For example, a periodic review of process hazards is generally appropriate. In addition, for workers who perform tasks or work in specific areas infrequently (e.g., a worker who spends 11 months per year in the shop but works in the process area during the annual shutdown), offering a review session for necessary tasks (i.e., just-in-time training) may be beneficial.

## 5.6 TRAINING FOR TECHNICAL PERSONNEL

Sometimes training discussions do not extend beyond the hourly workforce. However, because managers, supervisors, and engineers are also involved in the MI program, they must be competent and knowledgeable. Examples of the technical knowledge that may be necessary for supervisory and engineering personnel include (1) training in codes and standards and (2) training in the information needed to assess the suitability for degraded equipment to stay in service (e.g., corrosion mechanisms, fracture mechanics, stress analysis). Table 5-4 is a sample training plan for a position of Mechanical Integrity Engineer supporting a risk-based inspection (RBI) program for pressure vessels, storage tanks, and piping.

# TABLE 5-4
## Mechanical Engineer Training Plan

| Requirement | 1Y | 2Y | 3Y | 4Y | 5Y | 6Y | 7Y | 8Y |
|---|---|---|---|---|---|---|---|---|
| | \multicolumn{8}{c}{M = Mandatory; O = Optional} |
| Nondestructive examination (NDE) Limited Level II, ASNT SNT-TC-1A, Visual Testing (16 hrs) (waived if certified in API-510) | | | | | M | | M | |
| NDE Limited Level II, ASNT SNT-TC-1A, Liquid Penetrant (8 hrs) (waived if certified in API-510) | | | | O | | M | O | M |
| NDE Limited Level II, ASNT SNT-TC-1A, Radiographic Interpretation (24 hrs) (waived if certified in API-510) | | | | O | | M | O | M |
| NDE Limited Level II, ASNT SNT-TC-1A, Ultrasonic Straight Wave (8 hrs) (waived if certified in API-510) | | | | | M | | M | |
| Process Safety Management OSHA Process Safety Management (PSM) 1910.119 (24-40 hrs) | M | | | | | M | | |
| Mechanical Integrity (1910.119) (24-40 hrs) | | M | | | | O | | O |
| Fitness for Service – API 579 (24-40 hrs) | | | | M | | | O | |
| API-510 Pressure Vessel Inspection (certification optional) (24-40 hrs) | M | | | | | | O | |
| API-570 Piping Inspection (certification optional) (24-40 hrs) | | M | | | | | O | |
| API-653 Tank Inspection (certification optional) (24-40 hrs) | | | | M | | | | O |
| ASME Section VIII, Division 1, Design, Repair and Alteration of Pressure Vessels (40 hrs) | M | | | | | M | | |
| ASME B31.3 Process Piping (24-40 hrs) | | M | | | | | M | |
| ASME Section V, Nondestructive Testing (8-16 hrs) | | | | M | | | O | |
| ASME Section IX, Welding (16-24 hrs) | | | M | | | | | O |
| ASME/TEMA Design and Fabrication of Heat Exchangers (24-40 hrs) | | | | | | O | | M |
| EPA's 40 *Code of Federal Regulations* (CFR) Risk Management (24-40 hrs) | | | | | M | | | O |
| API-580/581 – Risk Based Inspection (24-40 hrs) | M | | | | | | | |
| Root Cause Failure Analysis (24-40 hrs) | | | | | | | M | |
| Chemical Engineering Concepts for Non-Chemical Engineers (24-40 hrs) | | | | | | | | O |
| Project Cost Estimating (24 hrs) | | | | | | O | | O |
| Basic Corrosion/Corrosion Concepts (24-40 hrs) | | | | | | | M | |
| Welding Inspection (24-40 hrs) | O | O | O | O | O | | | |
| National Fire Protection Association (NFPA) 70 Overview of the National Electric Code (NEC) (24-40 hrs) | | | | | O | O | O | O |
| Motor Protection (electrical) (40 hrs) | | | | | O | O | O | O |
| Pressure Vessel Design Software (24-40 hrs) | | | M | | | | | |
| RBI Software Training | M | | | | | | | |

# 5. MI TRAINING PROGRAM

## 5.7 CONTRACTOR ISSUES

Facilities hire contractors for a variety of activities: shutdowns and turnarounds, new construction, specialty services, and supplemental facility workforce. In some cases, all maintenance is contracted. In all cases, the facility is responsible for ensuring that contractors are trained in the skills and knowledge necessary to perform their tasks. Meeting this responsibility can involve directly training personnel, but more often it involves a review of the contractors' training programs and verification that each contract employee is trained. Facilities may have jurisdictionally required contractor safety programs with similar requirements for safety training (e.g., safe work practices, evacuation plans). Therefore, coordinating contractor oversight with the facility personnel who are administering contractor safety is important.

Any time that contract workers supplement a facility's workforce by performing tasks comparable to those performed by facility personnel, the facility staff should ensure that training is comparable. Measures to promote comparable training include (1) training contract workers directly or (2) providing contract employers with procedures and/or training materials. An alternative approach is to audit and approve contract employers' training programs and then verify that training is documented for each contract worker. Facilities for which all maintenance is contracted face similar choices for verifying that the training program is adequate and that all personnel are trained.

Often, contractors are hired to perform tasks that the facility cannot, or chooses not to, perform with facility personnel. In such cases, the facility may have no internal procedures to compare to the contractor training program; however, contractor training should still be reviewed. Certification or licensing is often required for specialty contractors (e.g., welders, inspectors, heavy equipment operators); facility personnel should verify that all necessary certifications and licensing are in place. For applicable contractors, the facility staff should also verify that all workers are trained and that the training is appropriately documented.

## 5.8 ROLES AND RESPONSIBILITIES

The roles and responsibilities for the training program are generally assigned to the Training, Engineering, and/or Maintenance departments. However, production personnel should provide guidance for topics such as process overviews and hazards. Example roles and responsibilities for an MI training program are shown in Table 5-5, presented in a matrix format that can be useful for many facility types. The matrix designates assignments to personnel with "R" as the person(s) responsible for the activity, "A" as the approver of the work or decisions made by the responsible party, "S" as persons supporting the responsible party in completing the activity, and "I" as persons informed when the activity is completed or delayed.

**TABLE 5-5**
**Example Roles and Responsibilities Matrix for the MI Training Program**

| Activity | Maintenance and Engineering Department Personnel ||||||| Other Personnel |||||
|---|---|---|---|---|---|---|---|---|---|---|---|---|
| | Maintenance Manager | Maintenance Supervisor | Maintenance Engineer | Project Managers/Engineers | Maintenance Technicians | Maintenance Planner/Scheduler | Production/Process Engineers | Process Safety Coordinator | Training Manager | Safety Manager | Plant Manager | Human Resource Personnel |
| **Skills/Knowledge Assessment** | | | | | | | | | | | | |
| • Identify job classifications | R | S | S | — | — | — | — | — | — | — | — | — |
| • Identify skills/knowledge requirements | A | R | R | S | S | — | — | — | — | S | — | S |
| • Survey the workforce | A | R | — | — | S | S | — | — | — | — | — | — |
| • Evaluate survey results and identify needs | A | R | S | S | — | — | — | — | — | S | — | — |
| **Training Program Development** | | | | | | | | | | | | |
| • Evaluate different approaches | A | R | S | — | S | — | — | S | S | S | — | — |
| • Obtain/develop training materials | S | S | S | — | — | — | — | — | R | S | — | S |
| • Establish documentation and tracking system | A | — | — | S | R | — | — | S | R | — | — | — |
| • Identify refresher training topics | A | S | R | S | — | — | — | — | — | — | — | S |
| • Establish certification programs | — | S | R | S | — | — | — | — | S | — | — | S |

## 5. MI TRAINING PROGRAM

**TABLE 5-5** *(Continued)*

| Activity | Maintenance and Engineering Department Personnel | | | | | | | Other Personnel | | | | |
|---|---|---|---|---|---|---|---|---|---|---|---|---|
| | Maintenance Manager | Maintenance Supervisor | Maintenance Engineer | Project Managers/Engineers | Maintenance Technicians | Maintenance Planner/Scheduler | Production/Process Engineers | Process Safety Coordinator | Training Manager | Safety Manager | Plant Manager | Human Resource Personnel |
| • Schedule training | — | S | — | | — | R | — | — | — | — | — | |
| • Establish ongoing training program | — | S | R | S | — | — | S | — | — | — | | |
| • Develop/implement technical training program | A | — | R | R | | | — | — | — | — | | |
| • Maintain training records | | S | S | S | — | S | | — | R | — | — | |
| • Develop Contractor Skills Evaluation Program | A | S | S | S | | | — | R | — | A | — | |

## 5.9 REFERENCES

5-1. ABSG Consulting Inc., *Mechanical Integrity, Course 111*, Process Safety Institute, Houston, TX, 2004.

5-2. American National Standards Institute, *Criteria for Accepted Practices in Safety, Health, and Environmental Training*, ANSI Z490.1, Washington, DC, 2001.

5-3. American Petroleum Institute, *Pressure Vessel Inspection Code: Maintenance Inspection, Rating, Repair and Alteration*, API 510, Washington, DC, 2003.

5-4. American Petroleum Institute, *Piping Inspection Code: Inspection, Repair, Alteration, and Rerating of In-service Piping*, API 570, Washington, DC, 2003.

5-5. American Petroleum Institute, *Tank Inspection, Repair, Alteration, and Reconstruction*, API 653, Washington, DC, 2001.

5-6. American Society of Mechanical Engineers, *International Boiler and Pressure Vessel Code*, New York, NY, 2004.

5-7. American Society of Non-destructive Testing (ASNT), *ASNT Central Certification Program*, ASNT Document CP-1, Columbus, OH, 2005.

5-8. American Welding Society, *Specification for Welding Procedure and Performance Qualification*, ANSI/AWS B2.1:2005, Miami, FL, 2005.

5-9. American Welding Society, *Specification for the Qualification of Welding Inspector Specialists and Welding Inspector Assistants*, AWS B5.2, Miami, FL, 2001.

5-10. National Board of Boiler and Pressure Vessel Inspectors, *National Board Inspection Code*, 2004 Edition, Columbus, OH, 2005.

# 5. MI TRAINING PROGRAM

## Appendix 5A. Sample Training Survey

Please complete the following survey to help management in its planned enhancement of our maintenance training program. The results of this survey will be combined with results of a similar survey of the maintenance foremen to establish priorities for the program.

This survey asks for an honest evaluation of your own knowledge and abilities. It also requests an honest general evaluation of your co-workers on the same topics. Thank you for helping us to improve the training program.

| Topic | Personal Capabilities | | | Co-worker Capabilities | | |
|---|---|---|---|---|---|---|
| | Very Capable | Some Knowledge | Basic Training Necessary | Very Capable | Some Knowledge | Basic Training Necessary |
| Pipefitting | | | | | | |
| Knowledge of piping specs | | | | | | |
| Knowledge of gasket selection | | | | | | |
| Bolt torquing | | | | | | |
| Crane operation | | | | | | |
| Forklift operation | | | | | | |
| Knowledge of emergency response procedures | | | | | | |
| Knowledge of the Alpha process | | | | | | |
| Knowledge of the Beta process | | | | | | |
| Knowledge of chemical hazards | | | | | | |

# 6
# MI PROGRAM PROCEDURES

An effective mechanical integrity (MI) program needs written procedures for MI program activities and specific tasks (e.g., inspection, testing, and preventive maintenance [ITPM] tasks). Written procedures help ensure that MI program activities and tasks are performed adequately, safely, and consistently, which are the primary reasons that process safety regulations include a requirement to develop and implement procedures for MI program activities (Reference 6-1). To develop and implement written procedures effectively, the facility staff should recognize that procedures offer value beyond compliance with regulatory requirements. Specifically, procedures can add value by:

- Serving as an integral part of the personnel training program
- Reducing human error
- Helping management ensure that activities are performed as intended
- Helping to provide continuity of the MI program during and after organizational and personnel changes

An important feature of successful MI programs is conformity among the written procedures, the corresponding training, and the actual field practice for a task or program activity. Differences among procedures, training, and practices can result in personnel confusion and mistakes. Therefore, to help ensure that consistent direction is provided, the information in the written procedures must be reflected in the training materials. In addition, management personnel should be directly involved in the MI procedure process by (1) providing resources needed to develop quality procedures, (2) ensuring that training resources are provided, and (3) monitoring procedure conformance.

Studies of major industrial incidents have shown that many of these events occur during or after maintenance activities (Reference 6-2). Human errors during the task/activity figure prominently in these events. Human errors can occur for many reasons, but procedure deficiencies are often contributing factors. These deficiencies can include:

- Lack of a current written procedure for the task/activity
- Discrepancies between the written procedure steps and actual training
- Failure to follow written procedures, usually because of (1) inadequate, missing, confusing, or incorrect information in the procedure, (2) a facility culture that does not support procedures as the accepted way to perform a task/activity (i.e., personnel are permitted to perform activities their way), (3) insufficient management systems (e.g., procedure audits) to ensure that procedures are followed, and/or (4) the existence of multiple conflicting procedures
- Lack of personnel training on the procedures

An effective program for developing and managing MI program procedures helps reduce human errors by ensuring that accurate procedures are established for needed tasks and that each procedure is appropriately written (e.g., simple clear statements, proper format, proper use of warnings). Developing and maintaining written procedures (i.e., keeping the procedure up to date) for every task and program activity is often impractical; therefore, management personnel should evaluate and prioritize tasks/activities to help ensure that the most critical procedures are developed and kept up to date. For example, a written procedure is often not necessary for low-risk, simple maintenance tasks, such as testing and changing alarm panel lights or tightening packing of valves in low-pressure, nonhazardous service. Also, studies related to human errors have found that employing simple writing techniques can effectively reduce human errors during maintenance tasks/activities. These writing techniques involve the use of easy-to-follow procedure formats, the correct placement and use of warnings and cautions, and the practice of writing instructions using clear, imperative statements.

Task and activity performance can be improved with well-written procedures that are accurate and user-friendly. Facilities should involve the personnel who will be performing the activities to (1) help prioritize the tasks/activities needing procedures and (2) help develop, validate, and maintain the written procedures.

The remainder of this chapter briefly discusses features of successful MI procedure programs. The Center for Chemical Process Safety (CCPS) book, *Guidelines for Writing Effective Operating and Maintenance Procedures*, provides more detail on developing and writing effective procedures. The following topics are discussed in the following sections of this chapter:

- Types of procedures supporting the MI program
- Identification of MI procedure needs
- Procedure development process
- MI procedure format and content
- Other sources of MI procedures
- Implementing and maintaining MI procedures
- Procedure program roles and responsibilities

# 6. MI PROGRAM PROCEDURES

## 6.1 TYPES OF PROCEDURES SUPPORTING THE MI PROGRAM

MI programs will often include different types of procedures. In general, the following types of procedures are found in an MI program (Reference 6-3):

- MI Program Procedures. These procedures outline the activities and the roles and responsibilities for the different elements of the MI program. In addition, documents that provide guidance or establish standards for MI activities, such as ITPM plans, inspection standards, and rationale for determining the equipment to include in the MI program, are part of MI program procedures. Program procedures for related programs that affect or are affected by the MI activities (e.g., management of change [MOC], process safety information) may be recognized and certainly will be referenced as part of the MI procedure program.
- Administrative Procedures. These procedures provide instructions for performing administrative tasks associated with the MI program.
- Quality Assurance (QA) Procedures. These procedures define the QA tasks to be performed and/or provide detailed instructions for performing QA tasks.
- Maintenance Procedures. These procedures provide instructions for performing core maintenance activities, job-specific or unique repair/replacement tasks, and equipment troubleshooting tasks.
- ITPM Procedures. These procedures provide instructions for performing ITPM tasks and recording/reacting to ITPM results.

In addition to the types of procedures listed above, safety procedures (e.g., lockout/tagout, line breaking, hot work permit, requirements for personal protective equipment [PPE]) are an important part of the MI program. These procedures are often referenced in various MI procedures (e.g., maintenance repair/replacement procedures, ITPM procedures) and are vital to helping ensure that MI tasks are performed safely.

How each facility categorizes the different types of MI procedures will vary; most facilities will have one or more procedures for each of the above-mentioned categories. In addition, the number of procedures needed for each category will also vary from facility to facility. Table 6-1 provides examples of the procedures that are typically developed by category. In addition, the CD accompanying this book includes example procedures for many of these categories.

In addition to the types of procedures mentioned above, procedures established by other departments may cover activities that are part of the MI program and should be recognized as MI procedures. Usually, such procedures include ITPM tasks or QA tasks that are performed by personnel outside the Maintenance or Inspection departments. For example, operating procedures (or portions of operating procedures) may provide instructions for such ITPM tasks as daily visual inspections of equipment, lubrication of pumps, maintenance of safety equipment in the process areas, testing of alarms, etc. Other departments that may have procedures related to the MI program are the Project Engineering, Environmental, Emergency Response, and Purchasing departments.

TABLE 6-1
Example MI Procedures

| Type of Procedure | Example Procedures | |
|---|---|---|
| MI Program | • MI program description<br>• Equipment selection and MI program applicability<br>• ITPM program development | • Equipment deficiency resolution<br>• Determining repair/replacement procedure needs<br>• ITPM plan/guidelines/standards |
| Maintenance, Repair/ Replacement Tasks | • Rupture disk replacement<br>• Hydrogen compressor mechanical seal replacement<br>• Large horsepower motor replacement | • Centrifugal pump disassembly and assembly<br>• Flammable solvent piping repairs<br>• Motor starter replacement<br>• Circuit breaker racking |
| Administrative | • Operation of the computerized maintenance management system (CMMS)<br>• Managing work orders<br>• Maintenance work planning<br>• Task schedule management | • Maintaining equipment files<br>• Reporting and coding equipment failures |
| QA | • Selecting and auditing contractors<br>• Selecting and auditing suppliers<br>• Project engineering technical and administrative activities<br>• Project engineering installation/ construction standards<br>• Technical and administrative activities for equipment repair or replacement | • Ordering, receiving, stocking, and issuing of spare parts and maintenance materials<br>• Decommissioning of equipment<br>• Equipment receipt inspection<br>• Shop-fabricated equipment inspection<br>• Positive material identification (PMI)<br>• Maintenance task oversight/review |
| ITPM Procedures (Note: Some regulations require written procedures for all ITPM tasks) | • Pressure vessel external and internal inspection<br>• Atmospheric storage tank periodic surveillance<br>• Atmospheric storage tank external and internal inspection<br>• Chemical cleaning of heat exchanger tubes<br>• Centrifugal pump vibration analysis<br>• Pump seal visual inspection<br>• Pump lubrication | • Relief valve removal and pop testing<br>• Conservation vent inspection and testing<br>• Interlock testing<br>• Transmitter calibration<br>• Instrument loop check<br>• On/off valve and position switch testing<br>• Fire sprinkler inspection<br>• Hose house inventory<br>• Tank farm dike visual inspection |

# 6. MI PROGRAM PROCEDURES

The developers of the MI program procedures should consider the different types of procedures needed, as well as the variety of users within a facility. The format, content, level of detail, and number of procedures will likely vary between different categories. For example, the format and content for a program procedure is usually quite different from that for an ITPM task because the frequency of use, the users' needs, and the detail required are different. For most facilities, more maintenance procedures are required than program or administrative procedures. Figure 6-1 groups the procedures into a common International Organization for Standardization (ISO)-9000 document tier structure and summarizes the general differences among the procedure types.

| Typical Number of Procedures | Specificity | Content | Typical Format |
|---|---|---|---|
| Few (e.g., 5 to 15) | Broad management-level issues | General instructions (what to do) (who does it) | Narrative Paragraph |
| Several (e.g., 10 to 30) | Specific scope with broad application | Mix of general and specific instructions | Paragraph Outline Checklists |
| Numerous (e.g., 20 to 200) | Specific scope (e.g., equipment, activities) | Specific instructions | Outline T-bar Checklists |

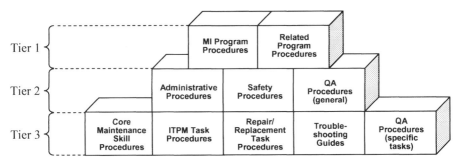

**FIGURE 6-1** MI procedure hierarchy.

The following two sections provide additional information on (1) determining what procedures are needed and (2) determining procedure content and selecting a good procedure format.

## 6.2 IDENTIFICATION OF MI PROCEDURE NEEDS

Determining the objectives of the procedure program is a good starting point for identifying procedure needs. The objectives can be determined by answering the following questions:

# 82 GUIDELINES FOR MECHANICAL INTEGRITY SYSTEMS

- What benefits (other than meeting regulatory requirements) are expected from the procedures? To train employees? To reduce human errors? To ensure program continuity as personnel change?
- What scope is to be included in the procedure program? Specific types of maintenance tasks? Specific units/processes/buildings?
- Who are the intended users for the different types of procedures?
- How are the different types of procedures to be used? Reference only? In the field each time the task is performed?
- Do applicable recognized and generally accepted good engineering practice(s) (RAGAGEPs) require a procedure for the activity?
- Are there regulatory requirements that must be met (i.e., for what tasks/ activities do the regulations require procedures)?

Assessing these issues helps a facility begin to understand the types of procedures needed, as well as some specific issues to be considered during procedure development.

Facility personnel should use a variety of information sources to compile a list of tasks and activities for each procedure type (e.g., MI program, ITPM procedures, QA procedures). Typically, the information can be found in the following:

- The survey used to assess training needs (see Section 5.1); alternatively, consider conducting a survey to assess procedure needs
- The applicable portions of the safety and environmental training manuals
- Any existing maintenance craft and job training protocol
- The ITPM plan (see Section 4.1 in Chapter 4)
- The QA plan (see Chapter 7)

Personnel can use these information sources as the primary tools for identifying the procedures needed for all of the procedure types except for repair/ replacement and troubleshooting task procedures. The procedure lists for repair/ replacement and troubleshooting tasks can often be developed by reviewing the work order history and/or interviewing the maintenance craftsmen and inspectors who perform these tasks.

Once compiled, the task lists must be evaluated to determine which activities warrant written procedures. To help improve employee buy-in, a team of personnel that includes one or more of the intended procedure users should participate in this evaluation. When performing the evaluation, the following factors should be considered:

- Is a written procedure needed to satisfy expected benefits (e.g., consistent execution of a task)?
- Is a written procedure necessary to ensure that the task/activity is executed correctly?

# 6. MI PROGRAM PROCEDURES

- Is the risk of incorrectly performing or not performing the task high enough to warrant a written procedure?
- Can the task be safely and correctly performed in only one way?
- Do the intended procedure users believe that a written procedure is needed for the task?
- Is the task more than a simple sequence of normal crafts skills?
- Will a checklist be useful for documenting completion of the task?
- Does a regulation (or RAGAGEP) explicitly require written procedures for these types of tasks/activities?

Evaluating most of these factors is usually straightforward and can be accomplished quickly. However, identifying higher-risk tasks and determining if a procedure is warranted can be more involved. For these tasks, expert judgment or the application of a simple risk-ranking tool can be used (Reference 6-4). (Note: Table 6-2 provides an example result from applying a simple risk-ranking tool.) The following questions should be considered when assessing risk:

- Who can be assigned to perform the task?
- What is the consequence if the task is not performed correctly?
- How often is the task performed?
- How likely will an error be made in performing the task that will result in the consequence?
- What safeguards (e.g., safe work practices) are in place?
- Is the risk of the task best managed by a procedure?
- How much training is required to best manage the risk?
- How effective is the training?

A risk assessment often results in a list of procedures to locate or develop for each procedure type (e.g., MI program, IPTM procedure, maintenance procedure). (In addition, Chapter 3 and Appendix C of the CCPS book, *Guidelines for Writing Effective Operating and Maintenance Procedures*, provide more information on determining which activities require written procedures [Reference 6-5]). The following section outlines a process for developing effective procedures.

## 6.3 PROCEDURE DEVELOPMENT PROCESS

An effective MI procedure program uses a structured procedure development process. A good procedure development process will help ensure that:

- The intended procedure users are involved, thus improving employee acceptance and ultimately improving the use of and compliance with procedures.
- Necessary information is incorporated by gathering information from a variety of sources.

**TABLE 6-2**
Example Risk Ranking Results for Procedure Determination

| Activity | Task Number | Activity Frequency | Risk Ranking ||| Procedure Required? | Comments |
| --- | --- | --- | --- | --- | --- | --- | --- |
| | | | Severity[1] | Likelihood[2] | Risk Ranking | | |
| Requesting materials and parts from stores | 1.1 | Daily | Significant | Occasional | Moderate | Yes | |
| Repairing and rebuilding pumps | 1.2 | Weekly | Major | Probable | High | Yes | |
| Repairing pump leaks | 1.3 | Weekly | Significant | Probable | Moderate | Yes | |
| Aligning pumps | 1.4 | Weekly | Significant | Probable | Moderate | Yes | |
| Repairing and rebuilding agitators | 1.5 | Once per year | Significant | Possible | Low | No | Team believes a procedure is needed |
| Repairing and overhauling compressor valves | 1.6 | 2 times per year | Catastrophic | Possible | High | No | Manufacturers' manuals are sufficient |
| Repairing and overhauling compressors | 1.7 | 2 times per year | Major | Possible | Moderate | Yes | |

[1] Severity ranking represents credible worst-case accident scenario if the task is incorrectly performed.
[2] Likelihood ranking considers both the frequency that the task is performed and the probability of the task being performed incorrectly, resulting in the credible worst-case accident scenario.

## 6. MI PROGRAM PROCEDURES

- Procedure information is presented in a logical and easy-to-use format by selecting an appropriate layout and following simple procedure-writing guidelines.
- Procedure content is accurate by using a multistaged review and validation process.

Figure 6-2 provides the basic steps of a good procedure-development process (Reference 6-6). At most facilities, the procedure development process rarely includes all of these individual steps. For example, many facilities combine review and approval into a single step, and the validation step is commonly not performed in a formal manner.

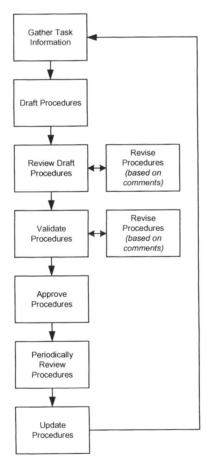

**FIGURE 6-2** Basic procedure development process.

The following paragraphs provide details and offer suggestions for improving the effectiveness of each step.

*Gather Task Information.* The objective of this step is to gather all of the information needed to draft the procedure. Necessary information includes (1) the main steps of the task/activity, (2) the hands-on steps (i.e., the discrete, physical actions needed to complete each main step), (3) the specific details related to the task (how, why, and who), (4) any special hazards involved in the procedure, and (5) special materials and tools, PPE, and necessary precautions. The procedure developer should obtain needed information by (1) interviewing subject matter experts (SMEs), (2) reviewing existing documents (e.g., procedures, manuals, drawings), and/or (3) observing the task being performed. During the information gathering step, the procedure developer (who may be an SME as well) should be careful to (1) not filter or screen out information as unimportant and (2) use open-ended questions when interviewing SMEs to avoid biasing the responses.

*Draft Procedures.* During this step, the information gathered in the previous step is converted into a user-friendly procedure. In doing this, the procedure developer determines which information to include in the procedure and where the information should be incorporated. Certain information is best placed in the introductory section of the procedure (e.g., hazards, special tools); some information belongs in the warning, caution, and note statements (e.g., hazard warnings, safety issues related to incorrectly performing a step); while other types of information should be placed directly in the procedure instruction steps (e.g., major steps, hands-on steps). Once these decisions have been made, the procedure developer incorporates the information into the procedure following an approved format and some simple procedure-writing guidelines (see Section 6.4 for additional information on procedure format and content).

*Review Draft Procedures.* The objective of the review step is to (1) verify that the approved procedure format has been followed, (2) confirm that the procedure meets regulatory (or other similar) requirements, (3) determine if the information provided by the SMEs and other sources has been appropriately incorporated into the procedure, and (4) assess the technical accuracy of the procedure. Because of its multiple objectives, this step usually involves numerous people in the organization. For example, safety department personnel may need to review certain procedures (e.g., ITPM procedures that involve confined space entry). At least one of the SMEs interviewed should be included in each procedure review. Establishing a formal review process helps to ensure that procedures are reviewed by the appropriate personnel. In addition, performing a proofreading review prior to submitting procedures for the full review process allows subsequent reviewers to focus on the technical aspects of the procedure rather than typographical and grammatical errors.

*Validate Procedures.* This procedure is intended to verify that the task can be executed as described in the procedure. The SME(s) who provided the information and the SME(s) involved in the procedure development to this point

# 6. MI PROGRAM PROCEDURES

often perform the validation step. In addition, SMEs with different experience levels may also participate in this review. The validation can be accomplished by performing a mock walk-through of the task as described or by following the procedure while actually performing the task. (Note: If the procedure is validated by actually performing the task, personnel performing the task must be made aware that the procedure has not been validated and may need to be altered.)

*Revise Procedures.* Procedures can be revised after the review step, after the validation step, and at any time it becomes necessary. The objective of this step is to address comments concerning the procedure. The procedure developer can resolve the comments by modifying the procedure to address the issue, or by convincing the commenter that the issue is already addressed in the procedure. If modified, the procedure should then receive an appropriate review before progressing to the next step.

*Approve Procedures.* The objective of this step is to obtain formal approval of the procedure from the appropriate person. If the review process is reliable, the approval step should proceed quickly.

*Periodically Review Procedures.* Effective procedures contain current and accurate information. To help maintain this accuracy, the procedure development process should contain a step requiring periodic review of procedures. Typically, this is accomplished by periodically resubmitting the procedure to a review process and/or by periodically observing the task while checking the procedure. Including the procedure users in the periodic review process can help ensure that the procedure reflects actual task steps and provides all of the necessary information.

*Update Procedures.* If deemed necessary (e.g., based on the periodic review), the procedures will be updated. The extent of the update will dictate which of the earlier procedure development steps to employ. For example, a simple update that involves expanding or correcting information for a couple of steps may not require a draft procedure review. On the other hand, a procedure that requires a complete rewrite may require all of the review steps. In addition, facilities must have a process (e.g., MOC process) to ensure that updated procedures are communicated to all personnel. When the procedure change is substantial, formal training on the changed procedure may be necessary.

Additional information on these development steps can be found in the CCPS book, *Guidelines for Writing Effective Operating and Maintenance Procedures* (Reference 6-5).

## 6.4 MI PROCEDURE FORMAT AND CONTENT

To create effective MI procedures, developers should use a format appropriate for the subject task and ensure that the content is appropriate. For the procedures to add value, they must both satisfy regulatory requirements (and other objectives, such as corporate mandates) and communicate task information in a manner that is clear, accurate, and easy to use.

Several factors influence the selection of the procedure format, including word processing capabilities and company requirements. However, the primary factor should be the type of task covered in the procedure. For example, MI program procedures are typically written in a narrative format (i.e., in paragraphs) or an outline format. On the other hand, ITPM task procedures are usually written using an outline format with short sections or a T-bar format (e.g., a two-column procedure format with a column containing the main steps and column providing the step details). (Note: The CD accompanying this book provides examples of common procedure formats, as well as advantages and disadvantages of the different formats. Chapter 4 and Appendix F in CCPS's *Guidelines for Writing Effective Operating and Maintenance Procedures* also provide additional information on the importance of procedure format and contain some sample procedure formats.) The format selected should:

- Be user-friendly
- Help simplify the procedure
- Help procedure developers input information in a consistent way
- Be appropriate for the type of task it is explaining

Another format consideration is the effective use of white space (on the page) and borders, and the selection of an appropriate type style and size. Benefits of using white space and borders effectively (especially for detailed, step-by-step maintenance task procedures) include:

- Uncluttered procedure presentation
- Logical grouping and/or separation of information
- Improved comprehension
- Improved readability

Ways to effectively create white space on a page include (1) increasing line spacing (e.g., use 1.5 or double spacing versus single spacing), (2) using indentation (e.g., indent substeps), or (3) presenting information in columns, tables, and/or bulleted lists. Any of these approaches can help prevent a discouraging "wall of words" appearance.

Similarly, the type style and size can enhance or detract from the procedure. Obviously, the type size must be large enough to read easily. The procedure developer can use uppercase words to emphasize important items. For example, if the standard use of the uppercase and lowercase is used, then items in all caps (e.g., WARNINGS, IF/THEN statements) will stand out and draw attention to information that is especially important.

Content, certainly, is the most essential ingredient of a procedure. Obviously, the information must be accurate and complete. The information must also provide the correct level of detail and be easy to use. Procedures that do not have

# 6. MI PROGRAM PROCEDURES

these attributes are typically not used and are of little value. This is especially true of detailed maintenance task procedures.

The user of a procedure should dictate the level of detail required. The procedure must provide enough detail so that the least-experienced worker is provided sufficient information to execute the task. On the other hand, veteran craftsmen may also use these procedures; for these individuals, too much or inappropriate detail is unnecessary. In addition, too much detail can make a procedure too long and difficult to use.

These are manageable issues, however. For every procedure, personnel should define and use an adequate task information gathering process (see Section 6.3) and employ simple, proven procedure-writing guidelines, such as:

- Each instruction is written as a command.
- The proper level of detail is used throughout the procedure.
- On average, only one action is implied per instruction step.
- The procedure indicates when instruction step sequence is important.
- The procedure is written using only *common* words (no 50-cent words).
- Only acronyms/abbreviations/jargon that aid in readers' understanding of the procedure are used.
- Each step is specific (i.e., there is no room left to guess/interpret).
- The procedure is free of steps that require in-your-head calculations.
- Any graphics used are for the users' benefit.
- The references have been used to the readers' advantage.
- Each page header contains the procedure issue date and page number.
- The procedure includes an indicator at the end to confirm that user is at the end of the procedure.

The CD accompanying this book contains a detailed procedure-writing checklist that outlines some of these simple guidelines. (Note: The guidelines in this checklist were developed specifically for detailed, step-by-step instructions for a maintenance task procedure; however, some of these guidelines can be applied to procedures covering other types of work. Also, Section 5.3 of the CCPS book, *Guidelines for Writing Effective Operating and Maintenance Procedures*, provides a more extensive list of guidelines, as well as an explanation of the guidelines.) These guidelines present writing techniques that can help improve the readers' understanding by:

- Presenting the information in clear and concise statements, starting each step with a command, having only one implied action per instruction step/substep, and ensuring appropriate use of referenced material (e.g., other procedures, original equipment manufacturer [OEM] manuals).
- Enhancing clarity by using common words and terminology (e.g., jargon, acronyms, abbreviations).

- Reducing interruptions (i.e., times that the user must put down the procedure) by providing sufficient details, being specific, and using graphics when helpful.

Procedures for many tasks should also address potential deviations that might occur during the task, such as (1) failing to obtain proper permits, (2) performing critical steps out of order, and (3) mistakenly replacing a part with a part of different materials of construction. Potential deviations are often addressed by including appropriate precaution, caution, and/or warning statements in the procedure. For some tasks, such as ITPM tasks, additional information (e.g., acceptance criteria, actions to take if criteria are not met) should be included or referenced.

## 6.5 OTHER SOURCES OF MI PROCEDURES

Facilities often use outside sources of procedures in addition to those developed by the facility. While many regulations require procedures for MI activities, the requirements rarely state that the facility must develop its own procedures for every task. Many companies have also determined that contractors performing a task or vendors supplying equipment are better able to develop procedures for some MI tasks.

Facilities cite two common reasons for using outside procedures (1) facility personnel lack the expertise or experience needed to develop the procedure and (2) a contractor or vendor already has the procedure available. In addition, OEM manuals may provide sufficient information to function as the procedure for some tasks. (Note: Many times, manufacturers' manuals lack the necessary permit and safety information needed to safely perform tasks in the field.)

Facility personnel should review any outside procedures that are used or referenced. The following are some questions to answer during such a review:

- Does the procedure include steps for the tasks to be performed?
- Is sufficient detail presented to ensure safe and consistent execution of the task(s)?
- Do the procedure details match the intended users' skill/knowledge base?
- Does the procedure contain necessary safety information?
- How well does the procedure interface with the facility's safety procedures (e.g., safe work practices)?
- Does the procedure contradict any facility safety procedures/practices?
- Does the procedure address facility quality or environmental concerns (e.g., product contamination, proper disposal of wastes)?

Facilities often find that they need to supplement contractor or supplier procedures to address site-specific issues or company policies.

## 6.6 IMPLEMENTING AND MAINTAINING MI PROCEDURES

The previous sections of this chapter have outlined many of the activities needed for an effective MI procedure program. However, two further phases of the procedure program are needed to ensure success of the program:

1. Implementation of the procedures
2. Maintenance of the procedures

While implementation may seem as simple as approving the procedures and then assembling them in manuals, successful implementation requires that the following items be addressed:

- Document control. Facilities with a document control system should include procedures in this system. Facilities should also take steps to dispose of old and out-of-date procedures when issuing the new documents.
- Access. Procedures should be readily accessible to employees. Employees are more likely to use procedures that are kept in a location that does not intimidate the employee. For example, procedures kept in a supervisor's office are less likely to be used by experienced employees if they must go to the supervisor to get a copy.
- Training. Procedure users should be trained on all new procedures to ensure that they fully understand the procedures. Training also provides an opportunity to verify that the procedures reflect how the work is actually performed.

Like physical assets, procedures can degrade with time (e.g., become inaccurate). Therefore, the procedure program should include the following methods for maintaining procedures:

- MOC. Changes to tasks covered by procedures must be managed for the procedures to remain accurate. This includes managing changes that result from improved ways to perform the task/activity, physical changes in the field (e.g., addition of new equipment), and/or organizational changes. Facilities can use existing MOC practices or develop a stand-alone change control process for the procedure program.
- Periodic review. Periodic reviews of MI procedures provide an effective means to ensure that the procedures are current and accurate and, therefore, more likely to be used. In some cases, periodic procedure review can be incorporated into refresher training activities.

Maintaining procedures requires management commitment; specifically, this requires that management provide the resources needed to update and revise procedures. Additional information on the topics discussed in this section can be found in the CCPS book, *Guidelines for Writing Effective Operating and Maintenance Procedures.*

## 6.7 PROCEDURE PROGRAM ROLES AND RESPONSIBILITIES

The roles and responsibilities for the MI procedure program can be assigned to personnel in various departments. Typically, the personnel in the Maintenance, Engineering, and EHS departments will be primarily involved. However, production personnel are likely to be involved in procedures for MI activities performed by production personnel. In addition, outside personnel (e.g., contractors) may also be involved. Example roles and responsibilities for an MI procedure program are shown in Table 6-3. The matrix designate assignments to personnel with "R" as the person(s) responsible for the activity, "A" as the approver of the work or decisions made by the responsible party, "S" as persons supporting the responsibility party in completing the activity, and "I" as persons informed when the activity is completed or delayed.

## 6.8 REFERENCES

6-1. Occupational Safety and Health Administration, *Process Safety Management of Highly Hazardous Chemicals*, 29 CFR Part 1910, Section 119, Washington, DC, 1992.

6-2. Marsh & McLennan, *Large Property Damage Losses in the Hydrocarbon Chemical Industries – A Thirty Year Review*, 20th Edition, New York, NY, 2003.

6-3. ABSG Consulting Inc., *Mechanical Integrity, Course 111*, Process Safety Institute, Houston, TX, 2004.

6-4. Brown, E. and R. Montgomery, *How to Determine the Training and Procedures Required to Support the Plant's Reliability Program*, presented at the MARCON 99 Conference, Gatlinburg, TN, 1999.

6-5. American Institute of Chemical Engineers, *Guidelines for Writing Effective Operating and Maintenance Procedures*, Center for Chemical Process Safety, New York, NY, 1996.

6-6. ABSG Consulting. Inc., *Writing Effective Maintenance Procedures, Course 114*, Process Safety Institute, Houston, TX, 2004.

## 6. MI PROGRAM PROCEDURES

**TABLE 6-3**
**Example Roles and Responsibilities Matrix for the MI Procedure Program**

| Activity | Inspection Manager | Maintenance Manager | Maintenance Engineers | Maintenance Supervisors | Inspectors | Maintenance Technicians | Maintenance Planner/Scheduler | Engineering Manager | Project Managers | Project Engineers | Area Superintendent/Unit Manager | Production/Process Engineers | Production Supervisors | Operators | Plant manager | EHS Manager | Process Safety Coordinator | Contractor | Equipment Vendor |
|---|---|---|---|---|---|---|---|---|---|---|---|---|---|---|---|---|---|---|---|
| **Reviewing and Validating Procedures** | | | | | | | | | | | | | | | | | | | |
| • MI program and related procedures | R | R | S | S | S | | | | | | | | | | | | S | | |
| • Administrative procedures | — | — | R | S | R | S | S | | | | | | | | | | S | | |
| • ITPM procedures | — | — | R | R | R | S | S | R | S | S | — | R | R | S | — | | — | S | S |
| • QA procedures | | | R | S | R | S | S | | | | | | | | | | — | S | S |
| • Maintenance procedures | | | S | R | | S | S | | | | | | | | | | — | | |
| **Approving Procedures** | | | | | | | | | | | | | | | | | | | |
| • MI program and related procedures | R/A | R/A | | | | | | | | | | | | | R/A | | R/A | | |
| • Administrative procedures | R/A | R/A | | | | | | | | | | | | | | | | | |
| • ITPM procedures | R/A | R/A | | | | | | R/A | | | R/A | | | | R/A | | | | |
| • QA procedures | R/A | R/A | | | | | | R/A | | | | | | | | | | | |
| • Maintenance procedures | | R/A | | | | | | | | | | | | | | | | | |

# TABLE 6-3 (Continued)

| Activity | Inspection Manager | Maintenance Manager | Maintenance Engineers | Maintenance Supervisors | Inspectors | Maintenance Technicians | Maintenance Planner/Scheduler | Engineering Manager | Project Managers | Project Engineers | Area Superintendent/Unit Manager | Production/Process Engineers | Production Supervisors | Operators | Plant manager | EHS Manager | Process Safety Coordinator | Contractor | Equipment Vendor |
|---|---|---|---|---|---|---|---|---|---|---|---|---|---|---|---|---|---|---|---|
| **Implementing Procedures** | | | | | | | | | | | | | | | | | | | |
| • MI program and related procedures | R | R | S | S | S | S | S | R | | | | | | | A | | S | | |
| • Administrative procedures | R | R | SI | S | S | S | S | | S | | | | | | A | | — | | |
| • ITPM procedures | A | A | S | R | R | S | S | R | | | R | S | S | | | | — | R | R |
| • QA procedures | A | A | S | R | R | | | | S | S | | | | | | | — | R | R |
| • Maintenance procedures | | A | S | R | | | | | | | | | | | | | — | | |
| **Maintaining Procedures** | | | | | | | | | | | | | | | | | | | |
| • MI program and related procedures | R | R | S | S | S | S | S | R | S | S | | | | | A | | S | | |
| • Administrative procedures | A | A | R | S | R | S | S | | | | R | S | S | | | | — | R | R |
| • ITPM procedures | A | A | S | R | R | S | S | A | | | | | | | | | — | R | |
| • QA procedures | A | A | S | R | R | S | S | | R | S | R | S | S | | | | — | R | R |
| • Maintenance procedures | | A | S | R | | S | S | | | | | | | | | | — | R | R |

# 7
# QUALITY ASSURANCE

A life-cycle approach to equipment quality assurance (QA) considers quality from the time the equipment is designed until the time it is taken out of service (for retirement or reuse). Because QA can be difficult for an external auditor to judge (most citations against QA programs are in response to accidents; few programs are deemed inadequate in advance of serious failures), some facilities may be tempted to pay little attention to a QA program. However, an effective QA program can be a powerful tool for upgrading a plant's entire mechanical integrity (MI) program.

In an MI program, QA and quality control (QC) work together to help ensure that appropriate tools, materials, and workmanship combine to provide equipment that performs to meet its design intentions. However, from company to company, the terms QA and QC can carry different connotations. In fact, the terms are often used interchangeably. Throughout this book, "quality assurance" or "QA" will include both QA and QC activities.

This chapter presents several suggestions for QA activities applicable to different phases of an equipment life cycle; however, most facilities are not likely to implement every suggestion. Also, depending on the importance of particular equipment, some QA activities may be more or less rigorous. For example, some facilities use positive material identification (PMI) only for specific processes or unique metallurgy.

A facility should examine existing practices at each stage of equipment life to determine whether QA deficiencies exist and, if so, develop a quality improvement plan to upgrade areas of vulnerability. In addition, facility personnel should develop a QA plan to be the basis of procedures and training for those activities. This chapter discusses QA activities for the following life-cycle stages:

- Design/engineering
- Procurement
- Fabrication
- Receiving
- Storage and retrieval

- Construction and installation
- In-service repairs, alterations, and rerating
- Temporary installations and temporary repairs
- Decommissioning/reuse
- Used equipment

In addition, this chapter discusses QA practices for spare parts and for contractor-supplied equipment and materials.

## 7.1 DESIGN

Design is usually the only opportunity a facility has to "build in" quality to its equipment; the remainder of QA is generally aimed at preserving quality. Quality designs start with competent and creative engineering. When possible, designs employ features that have been tested and proven; therefore, designs benefit from lessons learned so that mistakes are not repeated. Many of these proven designs then become the foundation for codes and standards and, more specifically, for company equipment specifications. All facilities should have equipment specifications. (Note that some companies refer to their own specifications as their design criteria.) Facilities without such documentation should add "develop equipment specifications/design criteria" to their quality improvement plans. Equipment specifications can begin with a reference to the codes and standards applicable to various equipment types. Table 7-1 contains a few widely used design codes. Another source of specifications for many processes are the design manuals or other information in the original engineering and purchasing records. In creating its own specifications, a facility using the original manuals and/or codes and standards should consider supplementing that information with (1) lessons learned that are specific to that facility (e.g., those documented during failure analysis investigations) and (2) updated or new information that may have become available after the original construction.

TABLE 7-1
Typical Design Code Applications

| Application | Design Code or Standard |
|---|---|
| Boilers (power) | ASME Boiler and Pressure Vessel Code (BPVC), Section I, Power Boilers |
| Electrical Systems | National Fire Protection Association (NFPA) 70 |
| Instrumentation | Various standards, including Instrumentation, Systems, and Automation Society (ISA) S84.01 |
| Piping (process) | ASME B31.3 |
| Pressure Vessels | ASME BPVC, Section VIII, Pressure Vessels |
| Pumps | Numerous standards, including API 610, ASME B73.1, and ASME B73.2 |
| Storage Tanks | API 620, API 650, and Underwriters Laboratories Inc. (UL) 142 |

# 7. QUALITY ASSURANCE

Companies should have systems in place to update specifications made necessary by (1) lessons learned from experience and investigations and (2) changes in the underlying codes and standards. The challenge of keeping specifications current is compounded following company mergers. Merged companies can be hampered by the use of "legacy specifications," leaving engineers with multiple versions of similar specifications. A similar problem occurs for companies that rely on vendor specifications when different vendors were used to build different process units. All companies should assign a person or team to address these issues and to be responsible for producing and maintaining a definitive set of specifications.

Maintaining proper design specifications is important. Documenting the design specifications and developing the supporting drawings and data sheets (i.e., process safety information) are often required by regulations. Applying the specifications correctly and having appropriate safeguards in place help produce a quality design. Companies use a variety of methods, including safety and design reviews (e.g., piping and instrumentation diagram [P&ID] reviews, relief system reviews, and various hazard reviews), to help ensure the quality of the designs. The Center for Chemical Process Safety (CCPS) publication, *Guidelines for Hazard Evaluation Procedures, Second Edition with Worked Examples*, includes descriptions of many hazard review techniques. The book also includes an extensive hazard evaluation checklist that many companies use to aid their design efforts. During the design stage, facility personnel should establish a QA plan for evaluating the design at the different design phases, as well as a QA plan to be used during fabrication and construction. Appendix 7A provides suggestions that can be incorporated into a QA plan for analyses during different phases of design (Reference 7-1).

## 7.2 PROCUREMENT

QA for procurement helps ensure that purchases adhere to a specified design, that change guidelines (i.e., knowing when substitution is acceptable and having appropriate approval for substitutions) are understood, and that qualified vendors are used. Often, QA is simpler if fewer people are involved: If parts and equipment purchasing can be carried out centrally (rather than having multiple departments involved) and only approved vendors are used, errors are less likely. All personnel involved in procurement should understand the specifications well enough to (1) recognize if inappropriate parts or materials are being ordered and (2) know when authorization is required for a material or part substitution.

Facilities should consider establishing a vendor approval procedure and limiting purchases to qualified vendors. Qualifying vendors helps to eliminate sources of improper parts and materials. A good vendor can benefit a plant's QA program by maintaining internal and external quality controls of its own. A sample vendor QA plan is in Appendix 7B (Reference 7-2).

## 7.3 FABRICATION

QA for fabrication includes verification that specifications are followed and that shop practices do not compromise quality. A vendor may be called upon to help with fabrication QA, but ultimately the facility is responsible for the quality of parts and equipment. Depending upon the importance of the equipment involved, facilities may use shop inspection and shop approval processes.

Shop and/or field fabrication site inspections are common QA tools. For pressure vessels and other critical equipment, the QA process often identifies hold points in the fabrication process. For example, before fabrication can continue, a company inspector or a third-party inspector may be required to verify the quality of root weld passes. Shop/site inspections may also be used more generally to oversee the fabrication procedures, shop/site conditions, and/or recordkeeping. To help make these inspections more effective, experienced inspectors and engineers should consider providing training, checklists, and/or procedures for others to follow. Some companies are using a shop qualification process to verify the procedures and QA practices of shops that may potentially fabricate their equipment. Such preapproval practices help ensure quality equipment fabrication, particularly if less-experienced personnel may perform project-specific inspections.

Many jurisdictions require using a code-approved shop for fabrication of some equipment (e.g., relief valves, pressure vessels). These shops have previously undergone an inspection and may continue to be inspected regularly by third parties (e.g., jurisdictionally authorized personnel). Some facilities have reported fabrication errors by these code shops, despite the authorizations they hold. Therefore, many facilities inspect and approve shops even when the shops already hold jurisdictional approval.

Several companies have also implemented some form of PMI. PMI has gained popularity with the realization that substandard, incorrectly shipped, and accidentally switched materials (e.g., stainless steel with lower-than-specified chromium content) have been sold to unsuspecting (and uninspecting) facilities. PMI is a process that facility procurement personnel can use to verify that the construction materials of a part or component fall within certain specifications. Companies and facilities implementing PMI perform some or all of the following activities: material testing, material tracking, and/or documentation tracking. In addition, the point at which companies execute PMI varies: some companies start with the steel mill run, others begin during fabrication, still others begin PMI during receiving. A facility may hire a third party to provide independent testing services, or company personnel may perform the testing themselves. In most programs, after the material composition has been positively identified, the results are documented and the component is tracked until installation. One approach for companies interested in PMI is to identify opportunities (e.g., projects with critical materials concerns) and include PMI steps (e.g., material tests, documentation tracking) in the project QA plan. More information on PMI is located in Appendix 7C (Reference 7-3).

## 7. QUALITY ASSURANCE

## 7.4 RECEIVING

To maintain thorough QA for receiving, a facility must recognize all of its receipt pathways. If all parts and equipment are received at a central storeroom, QA efforts would be focused there; however, some facilities may have multiple departments bringing in materials. In addition, the methods by which contractors receive materials vary considerably between facilities (see Section 7.12). If multiple pathways are involved, QA efforts at later stages (e.g., during equipment installation) may need to be intensified since no central source exists to verify proper receipt QA.

QA for receiving generally involves some kind of receipt inspection to verify that the parts received are the same as those that were designed and purchased. In addition, receipt inspection may be used to inspect parts for damages or nonconformities. The type of inspection and the personnel involved can vary with the type of equipment received, as well as with the facility culture and resources. In its simplest form, receipt inspections involve checking the packing list against the applicable purchase order. At its most complex, receipt can be a time for formal inspection, including material testing and identification. Many facilities' practices vary according to the importance of the equipment received. In such cases, thorough receiving procedures and/or training are important to help ensure that receiving personnel know when a formal inspection is required. Personnel performing receipt inspections (e.g., of spare parts) should receive appropriate training (e.g., to conduct visual inspections).

Many companies track nonconformities found during receipt QA and share this information with other company locations. This enables a company to more quickly identify quality issues, such as poor vendor performance.

## 7.5 STORAGE AND RETRIEVAL

Many of the QA considerations for storage are equipment specific (e.g., areas with controlled humidity and/or static electricity for electrical components, practices for storing pressure safety valves [PSVs] upright and turning motors). General considerations, such as binning, labeling, and inventory control measures (e.g., first-in, first-out; cycle count; reorder procedures), help to ensure that proper parts are available, are not confused with others, and do not spend too much of their useful life in storage. In some cases, facilities supplement these measures by creating segregated storage areas (e.g., for exotic metals). All of these QA considerations are relatively simple to manage with procedures and/or training for storeroom personnel.

Developing QA steps for routine retrieval of materials from the storeroom during hours that the storeroom is attended is quite straightforward. Facilities can use part-numbering systems and/or drawings, along with work orders, to provide control and tracking of issued materials. Such retrieval systems can be set up through the facility's computerized maintenance management system (CMMS) program; however, many facilities have successfully employed manual (i.e., noncomputerized) systems for parts checkout.

Nonstandard storage and retrieval systems can result in a more difficult QA process. Informal storage of spare parts and materials (e.g., at process units) can present opportunities for these materials to be misapplied. Further difficulties arise when one unit can "borrow" parts from another. Consider setting limits on the amount and types of parts that can be stored locally. Also note that unused parts (e.g., bolts, gaskets, small valves) returned to storage can lead to hazards when mistakes are made. Consider having controls (e.g., procedures, training) on parts return practices.

Few storerooms are staffed around the clock. Facilities that rely on the training of their storeroom personnel to ensure appropriate parts and materials retrieval should consider providing additional controls for the times that the storeroom is unattended. One common practice is to restrict storeroom access to selected individuals. In such cases, all selected individuals should receive appropriate training and have procedures in place to ensure that storeroom and retrieval QA is not compromised.

For some facilities, restricting parts retrieval for particular work order items is not practical. For example, an oil production field technician may have maintenance rounds that encompass many miles; requiring trips to a central storeroom for each work order is unreasonable. In such cases, the technician's truck can be treated as another storeroom, with appropriate procedures, training, and documentation of replacement parts required.

## 7.6 CONSTRUCTION AND INSTALLATION

Construction and installation are the last chance in the equipment life cycle to compensate for any QA vulnerabilities at earlier stages. Companies and facilities that do not correct vulnerabilities in the earlier stages of the life cycle should intensify QA for construction and installation. Of course, errors made during installation can nullify a program full of good practices up to that point. Ensure that controls are in place to prevent and/or detect installation errors (e.g., mixing low temperature valves with carbon steel valves, incorrect alignment of rotating equipment) before they lead to failures.

Construction and installation QA procedures should mandate the use of qualified personnel. For many companies, this entails auditing contractor performance as well as providing training for company personnel. Installation specifications can also help provide guidance to these personnel. Most companies have contractor approval processes; however, such processes are more often developed to ensure liability coverage and/or safety, rather than to guarantee contractor performance. Facility management should consider adding MI considerations to the contractor approval process. Appendix 7D contains a sample service contractor QA plan (Reference 7-4). QA activities help ensure that personnel performing construction and installation are qualified, and that their finished work is acceptable. As during the equipment fabrication stage, hold points are often used to provide independent inspection opportunities at key times during construction. Inspection and testing of the final installation are also very

## 7. QUALITY ASSURANCE

common. Particular tests may include hydrotesting of pressure equipment, instrument and interlock testing, and water runs. Using procedures and/or checklists can contribute to the consistency and quality of these activities.

One of the final opportunities to identify QA issues is during a pre-startup safety review (PSSR). Facilities should consider requiring a QA review as part of PSSR activities. During such a review, the installed equipment can be compared to the design documentation, and any project-specified installation requirements can be verified. Facilities should have a means to (1) document discrepancies between design and installation, (2) evaluate whether these discrepancies are tolerable (this evaluation can be similar to a change review process), (3) make necessary corrections prior to equipment startup, and (4) document closure of any identified items.

## 7.7 IN-SERVICE REPAIRS, ALTERATIONS, AND RERATING

Many in-service repairs occur in response to equipment deficiencies. General issues with equipment deficiency resolution are discussed in Chapter 8. Occasionally, facilities need to repair, alter, or rerate pressure vessels, tanks, and piping. (Repair is any work necessary to restore the equipment to a suitable state in accordance with the design conditions. Alteration is any physical change in equipment that has design implications, such as those changes affecting pressure-containing capabilities. Rerating is a change in the design temperature and/or the maximum allowable working pressure of the equipment.) Because of the potential catastrophic consequences of, and the technical issues involved with, this type of work, special QA requirements have been defined in applicable codes and standards. Table 7-2 lists some of the codes and standards applicable to repair, alteration, and rerating.

TABLE 7-2
Sample of Codes and Standards Having QA Requirements Applicable to Repair, Alteration, and Rerating

| Application | Applicable Code or Standard |
| --- | --- |
| Atmospheric Storage Tanks (i.e., API 650 tanks) | API 653, Section 7 |
| Electrical Systems | NFPA 70 |
| Instrumentation | ISA S84.01 |
| Low Pressure Tanks (i.e., API 620 tanks) | API 653, Section 7 |
| Piping | API 570, Section 8 |
| Pressure Vessels | API 510, Section 7 or National Board (NB)-23, part RC |

In general, these codes and standards provide requirements and guidance on the following issues:

- Authorization. The personnel who must authorize the work before it is performed.
- Approval. The personnel who need to approve the work once it has been performed.
- Workmanship. Details on specific aspects related to how the work is performed (e.g., welding techniques, heat treatment requirements, materials) and worker qualifications.
- Inspection and testing. Inspections and tests required during and after the work.
- Documentation. All of the documents that are required.

While having personnel who are experts on these issues is not imperative for a facility, it is important that organizations be aware of the issues and have access to a knowledgeable vessel/piping contractor who can assist with adhering to these requirements. Following these QA requirements will help to ensure the integrity of the equipment and avoid any legal or regulatory problems.

Construction and service history files can be helpful for personnel who are troubleshooting performance problems. In addition, records can be critical to safety for weld repairs. For example, post-weld heat treatment (PWHT) during original construction will generally dictate PWHT requirements for weld repairs. Catastrophic accidents during startup of vessels have occurred as a result of improper heat treatment. Some plants have developed work order review protocols intended to ensure that qualified personnel develop repair plans for critical repairs (e.g., repairs involving welding).

## 7.8 TEMPORARY INSTALLATIONS AND TEMPORARY REPAIRS

Specific problems may be presented by temporary installations and temporary repairs. Frequently, an installation or repair is designated "temporary" because it is not subject to the requirements of a permanent repair or installation. Sometimes, this includes bypassing QA checks required of permanent equipment. To help ensure that these situations do not lead to catastrophic consequences, facilities should consider implementing a policy for regulating temporary installations and repairs. Such a policy may be integrated into a facility's management of change (MOC) policy to help ensure that QA issues are adequately covered.

To manage temporary repairs and installations, facilities should ensure that these situations are identified and that any exceptions to specifications are noted. The situation should be reviewed to determine whether (1) modifications to operating limits are necessary for the duration of the installation, (2) the procedures should be updated, and (3) affected personnel should be informed of the changes. Depending on the type of temporary installation or repair, the equipment should be inspected prior to startup (restart), and additional inspections should be considered and scheduled as necessary. Finally, temporary installations/ repairs should be assigned an expiration date, and procedures should be developed to help ensure that the installation/repair is removed, upgraded, or rereviewed prior to that date.

## 7.9 DECOMMISSIONING/REUSE

QA for decommissioning is not an MI concern unless reuse is intended. Any equipment that is not removed and disposed of upon decommissioning is subject to reuse. "Mothballed" units and "boneyards" present opportunities for saving money, but they also present significant QA challenges.

Facilities that recognize these QA challenges should consider establishing decommissioning and recommissioning procedures. A decommissioning procedure should consider depressurization and cleaning of equipment, additional measures for equipment preservation, and any ongoing inspections and/or preventive maintenance (PM) that should be performed. In addition, design and inspection documentation should be retained, and equipment on its way to the boneyard should be labeled or tagged. Similarly, units that are mothballed for reuse at a later date (e.g., seasonally operated equipment) should have procedures to ensure that liquids are drained, systems are purged, and other measures are taken to help preserve equipment life (e.g., maintaining a proper atmosphere to prevent corrosion).

Some facilities provide QA for reusing equipment through recommissioning procedures. Recommissioning procedures may include a change-of-service approval process. Variables to consider in such an approval process include (1) the length of time the equipment was out of service and (2) the extent to which ongoing inspections and/or PM were performed. Generally, recommissioning involves inspections and other equipment checks to verify that the used equipment is suitable for the new service. If appropriate, pressure equipment may be rerated according to the practices listed in Section 7.7.

## 7.10 USED EQUIPMENT

Unforeseen problems can arise when purchasing used equipment. The purchasing facility may be provided design documentation, some data about previous equipment service, and/or previous repairs. Often, however, none of this information is available. Facilities that purchase used equipment should consider developing specific procedures for doing so. These procedures may include many of the same considerations listed in Section 7.9 for equipment reuse, such as recommissioning procedures. In addition, procedures for obtaining used equipment should identify methods for securing and maintaining equipment file information.

## 7.11 SPARE PARTS

Much of the information regarding QA of procurement, fabrication, receiving, storage, and retrieval systems (Sections 7.2 through 7.5) is directly applicable to managing spare parts. Additional considerations for spare parts management include (1) identifying spare parts to stock for new units/installations and (2) developing procedures and providing training for purchasing and receiving replacement parts and for approving substitute parts in place of original equipment manufacturer (OEM) parts.

## 7.12 CONTRACTOR-SUPPLIED EQUIPMENT AND MATERIALS

For facilities that allow contractors to supply their own equipment and/or materials, QA of those equipment/materials can either be covered by the facility's QA procedures or delegated to a contractor QA program. In either instance, facilities should ensure that all contractors are aware of the QA requirements. To the extent that contractors are providing QA services, facilities should consider auditing the contractors' practices. In addition, facilities should ensure that (1) all contractor-supplied equipment and materials are identified as soon as practical in the life cycle, (2) QA activities are initiated, and (3) equipment documentation is collected.

## 7.13 QA PROGRAM ROLES AND RESPONSIBILITIES

The roles and responsibilities for the QA program are generally assigned to Engineering, Inspection, and/or Maintenance departments. Example roles and responsibilities for the program are provided in Table 7-3. The matrix designates assignments to personnel with "R" as the person(s) responsible for the activity, "A" as the approver of the work or decisions made by the responsible party, "S" as persons supporting the responsible party in completing the activity, and "I" as persons informed when the activity is completed or delayed.

## 7.14 REFERENCES

7-1  Casada, M., R. Montgomery, and D. Walker, *Reliability-focused Design: Inherently More Reliable Processes Through Superior Engineering Design*, presented at the International Conference and Workshop on Reliability and Risk Management, San Antonio, TX, 1998.

7-2  ABSG Consulting Inc., *Mechanical Integrity, Course 111*, Process Safety Institute, Houston, TX, 2004.

7-3  American Petroleum Institute, *Material Verification Program for New and Existing Alloy Piping Systems*, API Recommended Practice (RP) 578, Washington, DC, 1999.

## 7. QUALITY ASSURANCE 105

## Additional Sources

American Institute of Chemical Engineers, *Guidelines for Hazard Evaluation Procedures, Second Edition with Worked Examples*, Center for Chemical Process Safety, New York, NY, 1992.

American Petroleum Institute, *Pressure Vessel Inspection Code: Maintenance Inspection, Rating, Repair and Alteration*, API 510, Washington, DC, 2003.

American Petroleum Institute, *Piping Inspection Code: Inspection, Repair, Alteration, and Rerating of In-service Piping*, API 570, Washington, DC, 2003.

American Petroleum Institute, *Centrifugal Pumps for Petroleum, Petrochemical and Natural Gas Industries*, API 610/ISO 13709, Washington, DC, 2004.

American Petroleum Institute, *Design and Construction of Large, Welded, Low-pressure Storage Tanks*, API 620, Washington, DC, 2002.

American Petroleum Institute, *Welded Steel Tanks for Oil Storage*, API 650, Washington, DC, 1998.

American Petroleum Institute, *Tank Inspection, Repair, Alteration, and Reconstruction*, API 653, Washington, DC, 2001.

American Society of Mechanical Engineers, *International Boiler and Pressure Vessel Code*, New York, NY, 2004.

American Society of Mechanical Engineers, *Process Piping*, ASME B31.3, New York, NY, 2004.

American Society of Mechanical Engineers, *Specification for Horizontal End Suction Centrifugal Pumps for Chemical Process*, ASME B73.1, New York, NY, 2001.

American Society of Mechanical Engineers, *Specifications for Vertical In-line Centrifugal Pumps for Chemical Process*, ASME B73.2, New York, NY, 2003.

American Society of Testing and Materials International, *Standard Guide for Metals Identification, Grade Verification, and Sorting*, ASTM E1476-97, West Conshohocken, PA, 1997.

The International Society for Measurement and Control, *Functional Safety: Safety Instrumented Systems for the Process Industry Sector - Part 1: Framework, Definitions, System, Hardware and Software Requirements*, ANSI/ISA-84.00.01-2004 Part 1 (IEC 61511-1 Mod), Research Triangle Park, NC, 2004.

National Board of Boiler and Pressure Vessel Inspectors, *National Board Inspection Code*, 2004 Edition, Columbus, OH, 2005.

National Fire Protection Association, *National Electrical Code*, NFPA 70, Quincy, MA, 2002.

Pipe Fabrication Institute, *Standard for Positive Material Identification of Piping Components Using Portable X-Ray Emission Type Equipment*, New York, NY, 2005.

Underwriters Laboratories Inc., *Steel Aboveground Tanks for Flammable and Combustible Liquids*, UL 142, Northbrook, IL, 2002.

**TABLE 7-3**
**Example Roles and Responsibilities Matrix for the QA Program**

| Activity | Maintenance and Engineering Department Personnel ||||||||| Other Personnel ||||||
|---|---|---|---|---|---|---|---|---|---|---|---|---|---|---|---|
| | Maintenance Manager | Engineering Manager | Maintenance Supervisors | Maintenance Engineers | Project Managers/Engineers | Maintenance Technicians | Maintenance Planner/Scheduler | Production/Process Engineers | Inspection Manager | Safety Manager/Process Safety Coordinator | Plant Manager | Inspectors | Storeroom Personnel | Purchasing Personnel | Contractor/Construction Supervisor |
| QA improvement plan development | A | A | S | R | R | – | S | S | A | – | – | – | – | – | – |
| QA plan development | A | A | S | R | R | – | S | S | A | – | – | – | – | – | – |
| Design QA (including specification development) | – | R | – | – | S | – | – | S | – | – | – | – | – | R | – |
| Purchasing QA | – | A | – | S | S | – | – | S | R | – | – | S | – | – | – |
| Fabrication QA | – | A | – | – | A | – | – | S | A | – | – | – | – | – | – |
| Receiving QA | – | – | R | A | A | – | – | A | A | – | – | – | R | – | – |
| Storage and retrieval QA | R | S | – | S | S | – | – | – | A | – | – | – | R | – | – |
| Construction and installation QA | S | R | – | – | S | – | – | – | A | – | – | – | – | – | – |
| Equipment rerating | R | – | – | S | – | – | – | – | A | – | – | – | – | – | – |
| Equipment repair QA | A | A | – | S | S | – | – | S | A | – | – | – | – | – | – |
| Temporary repairs and installations | A | A | – | S | S | – | – | – | – | R | – | – | – | – | – |
| Contractor QA | R | R | – | S | S | – | – | – | A | – | – | – | – | – | R |
| Decommissioning/recommissioning QA | – | – | – | S | S | – | – | – | A | – | – | – | – | – | – |
| Used equipment QA | S | R | – | S | S | – | – | – | A | – | – | – | – | S | R |

# 7. QUALITY ASSURANCE

**Appendix 7A. Design Review Suggestions**

The design process for a significant project has several steps (phases), and the QA activities may include different design reviews at each phase.

*Evaluation Phase.* Evaluation is the beginning phase of the design process. At this stage, some general information is available on the needs/requirements for a new process and the range of technologies that may be employed to satisfy those needs/requirements. The evaluation phase determines (1) whether this project should be pursued (relative to other projects competing for the same resources), (2) the specific objectives for the project, and (3) what fundamental direction (such as primary technological choices) the project should take. Analysis during this phase focuses on determining the feasibility of, and the risks associated with, the project. Much of the effort at the evaluation phase is in selecting which projects to pursue from among many competing alternatives — a decision in which the total cost of ownership (and hence, integrity-related characteristics) should play a key role.

*Conceptual Design Phase.* Conceptual design is an intermediate phase of the design process. During this phase, a development team investigates and proposes the "best" type of process configuration for the project. Choices about general types of processing equipment, interconnections between equipment, ranges of expected operating conditions, and operating environments are made during this phase; however, specific selection of components and process parameters will not be determined until later stages of the design process. Analysis during this phase focuses on determining how overall performance goals/objectives translate into goals/objectives for individual systems. This analysis provides an assessment (at a high level) of whether required performance is realistically achievable and what modifications/improvements would need to be made to attain overall performance goals/objectives. In addition, various design concepts are compared to determine which option(s) warrants further development based on a variety of factors, including project risk and expected life-cycle costs. Several analysis tools (e.g., what-if analysis, relative ranking, Pareto analysis) can be applied at this phase. Analyses at this stage strongly affect (support) equipment selection and configuration decisions.

*Preliminary Design Phase.* Preliminary design is another intermediate phase of the design process. During this phase, a development team expands on the work of the conceptual design phase by (1) finalizing choices about types and numbers of various equipment items, (2) producing the drawings that define the configuration of the equipment, (3) optimizing process parameter choices, and (4) evaluating impacts of tie-ins to plant services (e.g., air systems, electrical systems). Specific preplanning for how the process will be operated (such as operating procedure development) and how the process will be maintained (such as planned maintenance task/frequency definition) occurs in this phase. Analysis during this phase focuses on determining how individual system goals/objectives translate into component goals/objectives. Analysis also provides an assessment (at more detailed levels) of whether required performance of the individual systems is realistically achievable and what modifications/improvements to the individual systems or components are necessary. Optimization of system performance characteristics is completed during this phase and must consider a number of factors, such as costs, quality of products, and reliability-related characteristics (e.g., average system availability, system capability). Some analysis tools commonly used during this phase include:

- What-if analysis focusing on major items
- Relative ranking to evaluate key component reliabilities for competing designs
- Block diagram analysis to assess overall system reliability characteristics (e.g., overall equipment effectiveness) based on estimated or allocated individual system reliability characteristics

Analyses during this phase also support decisions regarding equipment selection and configuration, but generally go further to begin characterizing overall system performance. Issues important for reliable maintenance and operation, as well as manufacturing/assembling the process, begin to surface in these analyses.

***Detailed Design Phase.*** Detailed design is the final formal phase of the design process. During this phase, final choices are made regarding the specific equipment in the process (and associated vendors) and the layout of equipment. This phase could also include (1) completing the final design, (2) preparing equipment files, (3) developing and commissioning operating procedures/ instructions, (4) developing and scheduling inspection, testing, and preventive maintenance (ITPM) tasks for the process, and (5) selecting a spare parts stocking strategy for the process. The product of the detailed design phase should be a complete specification of equipment that can be successfully fabricated/installed, operated, and maintained. Analysis during this phase focuses on ensuring that equipment selection and configuration allow systems to meet design goals/objectives. An assessment is performed at the equipment level to determine whether required performance of the equipment (and ultimately the individual systems and overall system) is realistically achievable and, if not, what modifications and/or improvements are necessary. Optimization of equipment selection (based on factors such as equipment cost, quality of products, equipment reliability-related characteristics, etc.) is completed during this phase. In addition to equipment selection and configuration, analyses in this phase serve to identify:

- Major loss contributors for key systems (or components)
- Critical parameters for reliable fabrication/construction/ manufact-uring
- Important operating limits and startup criteria
- Appropriate planned maintenance tasks
- Necessary spare parts/materials stores

# 7. QUALITY ASSURANCE

The detailed design phase typically requires more analyses and tends to use a larger variety of analysis techniques than previous phases. Simpler techniques (e.g., checklists, what-if analysis) provide appropriate amounts of resolution to assess certain systems and particular equipment reliability, while other systems (e.g., complex, redundant systems) will require a more complex analysis technique, such as fault tree analysis (FTA) or common cause failure analysis. More detailed and more complex analyses are only applied to areas of significant risk or areas in which significant uncertainties exist regarding the level of risk. Never perform more analyses than are necessary for decision making. Chapter 11 discusses tools for risk-based decision making.

## Appendix 7B. Sample Vendor QA Plan

For major equipment items (e.g., a new pressure vessel) covered under the MI program, our facility may require that a vendor develop and implement a QA plan to ensure the quality of the item before use. The following vendor QA plan can be used as an audit tool for companies approving new vendors. A vendor QA plan may have (but is not limited to) the following features.

*Statement of Design and Fabrication Specifications.* For equipment that our personnel designed, this statement provides verification that the vendor adequately understands our specifications. For equipment designed by the vendor, this statement ensures that the vendor has formally established the specifications and allows for our input into the specifications.

*Definition and Scheduling of QA Activities.* This section defines the specific tasks that will promote/verify the quality of the equipment item. These tasks may include (as appropriate):

- Material quality verifications
- Worker qualification/training verifications
- Procedure quality verifications
- Fabrication specification verifications
- Manufacturing/fabrication process QC
- Nondestructive testing (NDT) during manufacturing/fabrication/field installation (dimensional checks; visual inspections; and hardness, ultrasonic, radiographic, performance, dye penetrant, pressure, and magnetic particle tests; etc.)

*Definition of Roles and Responsibilities for QA Activities.* This section establishes the roles of each organization (i.e., the vendor, our company, subcontractors to the vendor, and independent contractors) in accomplishing each of the required activities. The roles may include planning, performing, witnessing, evaluating, and/or documenting activities.

*List of Required Documentation for QA Activities.* This section lists the documentation that our company must receive to have confidence that the vendor has appropriately implemented the QA plan and that the quality of the equipment item is acceptable. This documentation may include (as appropriate):

- Manufacturers' data sheets (e.g., American Society of Mechanical Engineers [ASME] pressure vessel forms)
- Physical data sheets
- Mill test reports
- Material verification reports
- Weld maps
- Pressure test sheets

# 7. QUALITY ASSURANCE

- Reports for noted deficiencies and their resolutions
- Stress relief charts
- Hardness readings
- Charpy impact results
- Calculations
- NDT interpretation results
- As-built drawings
- Nameplate facsimiles
- Bill of materials

The vendor is responsible for developing the QA plan (when one is required by our company) and submitting the completed plan to us for comments. Our personnel responsible for purchasing the item will review the plan and provide comments to the vendor. The responsible company personnel will also ensure that the requirements of the plan are complete before our company places the equipment into service. This person also resolves any quality deficiencies with the vendor before placing the equipment into service and ensures that our company receives/files all required documentation.

### Appendix 7C. Positive Material Identification

The consequences can be devastating when incorrect assumptions are made about materials of construction or when materials are inadvertently substituted. Several serious incidents have occurred as a result of the substitution of incorrect piping materials. Many of these incidents were attributed to substituting carbon steel for chrome alloys. In one case, the catastrophic failure occurred shortly after the substitution. In another case, the catastrophic failure occurred 20 years later.

Diligent control over the materials of construction, including welding consumables, used in chemical processes can help prevent incidents and yield economic benefits as well. Such controls are commonly referred to as PMI. A comprehensive PMI effort should be integrated into the facility's hazard management system through a variety of practices. Material control begins with the material selection phase of the design process. The material selection process involves the economic consideration of the expected materials' performance in the anticipated exposure environment. PMI covers control of the material condition (i.e., mechanical, physical, and corrosion-resistant properties for the application) and verification of the material composition (i.e., the elemental constitution).

*Material Condition.* Heat treatment, mechanical working, or surface treatment (or a combination of these actions) can adversely affect a material's condition. The material's condition may create sensitivity to degradation with small changes in environmental conditions, thus creating equipment hazards.

Improper material condition is a common cause of a material-related equipment failure because of the large number of end-state conditions resulting from combinations of heat treating, forming, and fabrication methods. Material condition may create equipment hazards by increasing the sensitivity of the material to degradation at environmental conditions different from what was anticipated in the selection process. Examples of controls for material condition include (1) controlling grain size for high-temperature applications and for fatigue applications, (2) solution annealing for cold-formed austenitic materials to improve stress corrosion cracking resistance, and (3) tempering bolts for hydrogen sulfide services.

Material condition is generally not determined through analysis for composition and physical appearance. Equipment hazards caused by material condition are managed using QA testing during design, material specification, and manufacturing. Facilities sometimes have limited ability to measure material condition because the tests often require sampling for destructive testing or specialty technical analysis, such as hardness testing or metallographic examination. Generally, control of material condition is dependent on strong engineering and purchasing procedures to ensure that the correct material condition is specified, ordered, and marked and that the appropriate mill test or equipment fabricator documentation is provided with receipt of the materials.

# 7. QUALITY ASSURANCE

***Material Composition.*** Quantification of the material's composition or identification of a specific alloy can be accomplished using several methods available to most facilities. Typical composition identification methods include:

- Portable X-ray spectroscopy, which determines composition of elements found in most alloys and can provide estimates of specific alloy designation.
- Portable optical emissions spectroscopy, which determines composition of elements found in most alloys and can provide estimates of specific alloy designation. When used in an argon atmosphere, carbon content can be determined in both ferrous and austenitic materials.
- Access to commercial or company testing laboratory to perform compositional analysis.

Sorting or classifying materials within metallurgical types can be accomplished using a variety of methods available to most facilities. Typical classification methods include:

- Classification or sorting using basic principles
  - Color – copper alloys versus white metals
  - Density – aluminum versus magnesium, titanium versus steel
  - Magnetic response – strong, weak, or no magnetic response to identify ferrous metals versus austenitic or nickel-based materials
- Chemical etches can determine some metals present and some types of surface coatings; care is required when handling reagents
- Resistivity testing to determine broad types of materials, but not alloy type
- Grinding wheel spark test – an experienced operator can differentiate ferrous materials and determine if the metal can be welded based on carbon content

These methods may be specified as part of the routine QA procedures or performed on an as-needed basis (e.g., for poorly labeled filler metal composition or for service materials without readily discernable markings).

Recognized and generally accepted good engineering practice (RAGAGEP) guidance for establishing formal material verification procedures as part of the facility's overall PMI practices is found in:

- API RP 578 – *Material Verification Program for New and Existing Alloy Piping*
- American Society of Testing and Materials (ASTM) E1476-97 – *Standard Guide for Metals Identification, Grade Verification, and Sorting Systems*

These documents provide guidance for establishing the basic elements of a material verification work process that includes (1) components to test and level of examination, (2) test methods, (3) acceptance criteria, (4) material marking, (5) test documentation, and (6) resolution of material nonconformances.

In addition, these documents suggest good practices to justify variances, including when specific material verification of all incoming material is generally not required. This may be acceptable if mill test report documentation is provided at material receipt in conjunction with physical verification of material markings for a representative sample. The documents suggest that the need for stringent or statistically founded composition verification increases for both new and existing materials when:

- Verification of a specific alloy grade is needed, or the specificity of the material's composition (such as for trace elements) is essential for performance.
- The cost of the material is high.
- An incident or near miss lists "incorrect material" as a causal factor.
- The facility has had unfavorable experience (or lack of experience) with a supplier, or the material is used infrequently at the facility.
- The consequences of failure are high.
- Historical fabrication and installation practices or facility documentation suggest poor material controls and/or poor material traceability.

*PMI Practices.* PMI practices, in addition to composition verification, are performed to help ensure that correct materials of construction, including welding consumables, are specified and used. PMI practices are applicable during any MI activity and are intended to complement the materials verification procedures. PMI practices include procedural controls and employee awareness activities that are usually developed and maintained by technical personnel within the facility. PMI helps to ensure the correct selection, purchase, receipt, and installation of process equipment with the appropriate materials of construction. Typical PMI procedures and practices for a facility may include:

- Equipment Design Standards
  - Materials selection and fluid compatibility guidelines
- Project Engineering Procedures
  - Shop inspection
  - Component inspection and material verification
- Process Safety Management
  - Process safety information
  - Process hazard analysis (PHA)
  - MOC
  - PSSR
  - QA procedures
- Maintenance Work Controls
  - Maintenance job planning to ensure replacement in kind or MOC approval for a change
- Engineering Standards or Procedures for Pressure Equipment
  - QA manual (e.g., ASME fabrication or repair)

# 7. QUALITY ASSURANCE 115

- Company engineering standards
- Employee Awareness
  - Knowledge of standard material markings (color coding and ASTM designations)
  - Knowledge of fabrication codes and standards
  - Basic materials incompatibility knowledge (e.g., stainless steel with chloride solutions, brass alloys with ammonia, aluminum alloys with high pH, carbon steel in strong acid/water solutions)
  - "Something's not right" awareness – appearance, nonstandard markings, weldability, or dimensional anomalies
- Weld Procedures and Weld Qualification Procedures
  - Technical approval of the weld procedure for each job
  - Control of filler metal identification and storage conditions
- Purchasing Controls
  - Order placement requires using recognized and specific material designations, such as the Unified Numbering System
  - Requirements for inclusion of appropriate documentation with receipt of materials
  - Technical approval requirements for ordering alloy or specialty materials
- Warehouse Controls
  - Receiving
  Confirmation of material markings with purchase order
  Confirmation of receipt of documentation, such as mill test reports
  - Storage
  Segregation by alloy
  Color coding or other markings
  Control of environmental conditions that could lead to degradation (chloride or sulfur compounds in the environment, standing water, etc.)
  - Issuing
  Issue matched to correct work order or project documentation
  Return-to-stock procedures for unused materials

## Appendix 7D. Sample Service Contractor QA Plan

*Service Contractor QA Plans.* For extensive maintenance, inspection and testing, and construction activities by service contractors (e.g., a significant process modification during a turnaround) within areas of the plant covered by the MI program, our company may require a contractor to develop and implement a QA plan to ensure the quality of their work. The following contractor QA plan can be used as an audit tool for companies approving new contractors. A contractor QA plan may have (but is not limited to) the following features.

*Statement of the Scope of Work.* For work planned by our company, this statement provides verification that the contractor adequately understands our objectives. For work planned by the contractor, this statement ensures that the contractor has formally established plans for their work and allows for our company input into these plans.

*Definition and Scheduling of QA Activities.* This section defines the specific tasks that will promote/verify the quality of the work. These tasks may include (as appropriate):

- Material quality verifications
- Worker qualification/training verifications
- Procedure quality verifications
- Fabrication specification verifications
- Fabrication process QC
- NDT during fabrication/field installation (dimensional checks; visual inspections; and hardness, ultrasonic, radiographic, performance, dye penetrant, pressure, and magnetic particle tests; etc.)
- Oversight of responsibilities by qualified supervisors

*Definition of Roles and Responsibilities for QA Activities.* This section establishes the roles of each organization (i.e., the contractor, our company, subcontractors to the prime contractor, and independent contractors) in accomplishing each of the required activities. The roles may include planning, performing, witnessing, evaluating, and/or documenting activities.

*List of Required Documentation for QA Activities.* This section lists the documentation that our company must receive to have confidence that the contractor has appropriately implemented the QA plan and that the quality of the work is acceptable. This documentation may include (as appropriate):

- Documentation of QA activities for specific equipment items
- Work completion checklists, including mechanical completion (e.g., all components and protective coatings/linings installed), electrical completion (e.g., all circuits and grounding completed), and

# 7. QUALITY ASSURANCE

instrumentation completion (e.g., control loops calibrated and interlocks tested)
- Material verification reports
- System integrity and performance test reports
- Reports for noted deficiencies and their resolutions
- Stress relief charts
- Calculations
- NDT interpretation results
- As-built drawings

The contractor (with input from all subcontractors) is responsible for developing the QA plan (when one is required by our company) and submitting the completed plan to us for comments. Our personnel responsible for hiring the contractor will review the plan and provide comments to the contractor. Our responsible personnel will also ensure that the requirements of the plan are complete before our company places newly installed/modified equipment into service. This person also resolves any quality deficiencies with the contractor before placing the new or modified equipment into service and ensures that our company receives/files all required documentation.

# 8
# EQUIPMENT DEFICIENCY MANAGEMENT

Successful mechanical integrity (MI) programs include effective plans for recognizing and reacting to equipment deficiencies. A deficiency is identified through the evaluation of equipment condition based on MI activity results or by the observation of substandard equipment performance or condition during normal operations. A deficiency is noted when an observed condition is outside the established limits (acceptance criteria) that define equipment integrity.

Deficient equipment conditions can be discovered (1) during acceptance testing for new equipment fabrication or installation, (2) in the course of performing inspection, testing, and preventive maintenance (ITPM) activities, or (3) while measurements are taken when the equipment is accessible during a repair. In addition, operations personnel can observe deficiencies that first appear as operating difficulties (the start of the traditional work order process). Without routine equipment assessments and subsequent evaluations of equipment condition, deficient conditions would remain unknown. Each MI activity should include work processes and procedures for making observations, performing the appropriate evaluations, and documenting the observations and evaluations. Likewise, operations personnel should have a procedure for documenting and reporting suspected equipment deficiencies. Equipment assessments and subsequent evaluations form the basis for determining whether the process can continue functioning (temporarily) while the deficient condition persists.

## 8.1 EQUIPMENT DEFICIENCY MANAGEMENT PROCESS

To effectively manage deficient equipment conditions, facilities must implement a process to ensure that the following actions occur:

- Acceptance criteria are established that define proper equipment performance/conditions
- Equipment condition is routinely evaluated

- Deficient conditions are identified
- Proper responses to deficient conditions are developed and implemented
- Equipment deficiencies are communicated to affected personnel
- Deficient conditions are appropriately resolved

The following sections discuss each of these actions, except for ensuring equipment conditions are routinely evaluated. This topic has been discussed in detail in Chapters 4 and 7. Chapter 4 discussed routine evaluation of equipment condition through ITPM activities. Chapter 7 discussed establishing quality assurance (QA) activities to evaluate equipment conditions throughout the life of the equipment. In addition to these actions, organizations should consider using equipment failure and/or root cause analysis (RCA) techniques to identify underlying causes of equipment deficiencies. (Chapter 12 contains more information on equipment failure analysis and RCA.)

## 8.2 ACCEPTANCE CRITERIA

Acceptance criteria are established to provide (1) confidence in the equipment's integrity while the deficient condition is being resolved (i.e., during the time required to implement permanent corrective actions), accounting for any uncertainty in the equipment assessment process; (2) confirmation of function; and/or (3) standards for new equipment fabrication, installation, or repair activities. Acceptance criteria are specific to each type of equipment and each observation method. An equipment condition that does not satisfy an acceptance criterion should not signal an immediate threat for catastrophic consequences; rather, the equipment's condition should indicate that attention from the integrity management system is needed.

Figure 8-1 illustrates the concept for selecting acceptance criteria for equipment conditions observed of in-service equipment. The acceptance criteria values should be conservative enough to provide time for further condition assessment and to accommodate uncertainty in the absolute condition of the equipment (i.e., the observation process may not reveal the worst condition of the equipment).

Equipment acceptance criteria are based on recognized and generally accepted good engineering practice(s) (RAGAGEPs) and/or on engineering calculations used as the design basis for the equipment. Such engineering calculations should be documented in the facility's process safety information. In addition, subject matter experts (SMEs) (e.g., inspectors, mechanics, equipment vendors) can be consulted and information from several resources can be used during the development of acceptance criteria. Table 8-1 provides examples of resources useful for determining acceptance criteria. In addition, Chapter 9 contains lists of RAGAGEPs applicable to many common types of process equipment.

# 8. EQUIPMENT DEFICIENCY MANAGEMENT

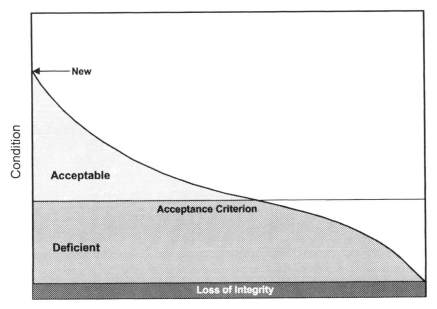

**FIGURE 8-1** Technical evaluation condition selection.

| TABLE 8-1 Acceptance Criteria Resources | |
|---|---|
| Vessels and Tanks | • Inspection codes and standards<br>• As-built drawings<br>• Vessel/tank specifications |
| Piping | • Inspection codes and standards<br>• Piping specifications |
| Instrumentation | • Manufacturers' manuals<br>• Instrument specifications<br>• Professional society (e.g., Instrumentation, Systems, and Automation Society [ISA]) documents |
| Rotating Equipment | • Manufacturers' manuals<br>• As-built drawings<br>• Equipment specifications |
| General | • Manufacturers' manuals<br>• Equipment specifications |

When determining acceptance criteria, personnel must consider a variety of potential equipment deficiencies that can occur and develop corresponding acceptance criteria based on measurable equipment conditions/attributes. Table 8-2 provides broad categories of equipment conditions/attributes that are commonly used to define acceptance criteria. The criteria can be defined quantitatively or qualitatively. The information is this table is presented for major types of equipment and three MI activity areas (1) new equipment design/fabrication/installation, (2) inspection and testing, and (3) repair.

These acceptance criteria are used when evaluating an observed condition to determine if the condition represents a deficiency requiring resolution. Developing all acceptance criteria prior to placing equipment into service may not be necessary, but limits must be known at the time of observation (e.g., inspection), or soon after, so that a detailed evaluation can be completed. Each acceptance criterion should also explain the actions required when the equipment's condition is found to exceed that criterion. For example, when pressure-containing equipment is determined to be too thin, the options often include rerating, repair, or replacement of the vessel, and/or performing fitness for service (FFS) analysis to set new acceptance criteria. (Appendix 8A contains an overview of the American Petroleum Institute's [API's] FFS recommended practice, API Recommended Practice [RP] 579 [Reference 8-1].)

## 8.3 EQUIPMENT DEFICIENCY IDENTIFICATION

Personnel may identify deficiencies during the performance of MI activities, either while the equipment is in service or out of service, or prior to the equipment being placed into service. Facility response to deficient conditions is time-dependent. The urgency of the response depends on (1) whether the condition is identified while the equipment is in service, (2) how close the condition is to loss of integrity (see Figure 8-1), and/or (3) the increase in risk to personnel or assets from the loss of function of a safety system. For equipment not in service, regulatory requirements and good common practices dictate that a deficiency be corrected before use.

To help ensure that deficiencies are identified, an MI program needs to include procedures for evaluating ITPM results. For each activity, the acceptance criteria and associated actions (such as further evaluation, communication, and resolution) should be included in the written procedure (see Section 6.4) to support the evaluation of the observation results (e.g., QA inspection, ITPM task results) and ensure that the equipment deficiency is properly managed.

Sometimes personnel with specific technical expertise may be required to determine (1) if equipment is deficient, (2) the appropriate precautions for continued operation, and (3) the ultimate resolution of the deficiency. Typically, specific expertise and appropriate methods are needed to evaluate more complex issues (e.g., localized corrosion, higher-than-expected corrosion rates, suspect welds) related to pressure-containing equipment (e.g., pressure vessels, tanks,

## 8. EQUIPMENT DEFICIENCY MANAGEMENT

piping). In addition, specific expertise and appropriate techniques are needed to perform FFS evaluations.

## 8.4 RESPONDING TO EQUIPMENT DEFICIENCIES

Operational and safety risks often accompany the following actions (1) continuing to operate equipment with a deficient condition, (2) placing equipment into service with a deficient condition, and (3) immediately shutting down equipment with a deficient condition. Therefore, facility management (ideally with the support of technical assurance personnel) should be responsible for these decisions. In addition, facilities should have written procedures in place and the appropriate means for ensuring that all hazards associated with the selected course of action are addressed and mitigated. The written procedures for deficiency resolution should include the following:

- Methods of hazard and risk assessment
- Management and engineering approval authority for temporary mitigation and corrective actions
- Tracking and closure of temporary mitigation measures
- Communication of hazards to affected parties
- Training, as necessary, for affected parties on temporary mitigation measures
- Equipment-specific documentation

The risks associated with deficiency resolution are best assessed with engineering and managerial perspectives. The level of detail of the risk assessment should (1) be dictated in the written procedure by the complexity and estimated risks associated with the deficiency and (2) provide the basis for establishing the timing and response for both the temporary mitigation measures and the ultimate corrective actions for the deficient condition. The response to an equipment deficiency occurs in stages, beginning with the definition of the problem (i.e., the identification and evaluation of a specific deficiency), through implementation of temporary mitigation measures, and culminating with the final corrective action. The facility's management of change (MOC) and pre-startup safety review (PSSR) systems can be used to document the stages and activities of the response path.

## TABLE 8-2
## Examples of Acceptance Criteria for Common Types of Equipment

| Equipment | New Equipment Fabrication and Installation | Inspection and Testing | Repair |
|---|---|---|---|
| Pressure Vessels, Tanks, and Piping | • Pressure rating for design and testing<br>• Weld quality<br>• Dimensional/alignment tolerances<br>• Materials of construction<br>• Valve leakage rates<br>• Installation criteria for supports, pipe, and hangers | • Thickness requirements for each pressure boundary component<br>• Requirements for assessment of support and anchoring systems<br>• Foundation settlement limits<br>• Tolerance to linear indications<br>• Tolerance to distortion<br>• Leakage | • Weld quality<br>• Materials of construction<br>• Dimensional/alignment tolerances<br>• Pressure rating for design and testing<br>• Leakage<br>• Installation criteria for supports, pipe, and hangers |
| Pressure Relief Valves (PRVs) | • Materials of construction<br>• Design pressure and temperature<br>• Relief capacity<br>• Storage conditions<br>• Installation criteria<br>• Leakage | • Set testing pressure limits<br>• Pre-disassembly test limits<br>• Set pressure and blowdown tolerances<br>• Visual inspection (pipe and PRV) for fouling/plugging<br>• Leakage | • Materials of construction<br>• Dimensional tolerances<br>• Installation criteria<br>• Leakage |
| Instrumentation | • Materials of construction<br>• Installation criteria<br>• Component calibration<br>• Functional performance criteria | • Component calibration<br>• Functional performance criteria | • Materials of construction<br>• Installation criteria<br>• Component calibration |
| Rotating Equipment | • Materials of construction<br>• Performance testing criteria<br>• Pressure testing requirements<br>• Storage conditions<br>• Installation criteria<br>• Leakage | • Vibration limits<br>• Bearing or coolant temperatures<br>• Trip speeds<br>• Requirements for assessment of support and anchoring systems<br>• Leakage | • Weld quality<br>• Materials of construction<br>• Dimensional/alignment tolerances<br>• Pressure testing requirements<br>• Installation criteria<br>• Leakage |

## 8. EQUIPMENT DEFICIENCY MANAGEMENT

### TABLE 8-2 (Continued)

| Equipment | New Equipment Fabrication and Installation | Inspection and Testing | Repair |
|---|---|---|---|
| Fire Protection | • Installation criteria<br>• Pressure testing requirements<br>• Performance testing criteria | • Performance testing criteria<br>• Requirements for assessment of support and anchoring systems | • Installation criteria<br>• Weld quality<br>• Materials of construction<br>• Dimensional/alignment tolerances |
| Electrical | • Conformance to applicable National Electric Code (NEC) requirements | • Breaker testing performance | • Conformance to applicable NEC requirements |
| Fired Heaters | • Installation criteria for supports, pipe hangers, and burners<br>• Weld quality<br>• Materials of construction for components: tubes, supports, refractory and instrument taps | • Tube and instrument tap dimensions and condition<br>• Pipe hanger spring settings<br>• Refractory condition | • Installation criteria for supports, pipe hangers, and burners<br>• Weld quality<br>• Materials of construction for components, tubes, supports, refractory and instrument taps |

Typically, resolving deficiencies for in-service equipment involves determining if the equipment should be shut down. When opting to keep deficient equipment in service, facility personnel must be able to demonstrate that continued operation is safe (i.e., that the risk of operating with the deficiency is tolerable). In addition, personnel should document those factors that the decision to continue operation was based on. If a decision is made to leave the equipment in service, a variety of resolution paths are possible to help ensure continued safe and reliable operation (until the equipment is renewed or permanently repaired):

- Continued operation with adjustment of the ITPM method (e.g., increase in number/extent of thickness measurements, using different nondestructive examination [NDE] techniques) or schedule (e.g., more frequent monitoring of the degradation rate and extent of damage)
- Continued operation with (1) installation of temporary mitigating measures to provide replacement of lost function or to enhance integrity or (2) temporary changes to the operating conditions to minimize degradation rate
- Continued operation until the scheduled repair of the deficient condition or renewal of the equipment for permanent correction can be made. (Until that time, continued operation should follow technical recommendations to minimize degradation while making the equipment available at a specific time)
- Continued operation after an FFS evaluation; more details on FFS are included in Appendix 8A

The resolution of a deficient condition for out-of-service equipment involves a similar thought process. Facility management must decide if the equipment should be placed (back) into service before the deficiency is corrected. If so, similar resolution paths are appropriate to help ensure safe, reliable operation. Following the risk assessment, the resolution path involving a deficient condition for out-of-service equipment comprises at least one of the above in-service paths and one of the following paths:

- Immediate repair or renewal for permanent correction of the deficient condition
- Delayed startup until an evaluation to confirm FFS is completed and, if necessary, the scope of repairs or other response paths is defined

Temporary measures for some types of deficiencies may include clamps or other forms of leak sealing, new piping bypasses, and/or alternate process lineups for relief protection. Employing temporary mitigation measures and/or other temporary changes (e.g., safety systems placed on bypass, alternate fire protection or gas detection measures, operation without spare equipment or with temporary equipment) requires appropriate technical review and documentation, and includes a schedule for removal of the temporary measures or technical reevaluation. Such technical review and/or reevaluation can be considered part of the deficiency resolution process and should be tracked until the temporary measure is removed. The identification of an out-of-service deficient condition should be resolved using QA procedures and should be documented in the equipment information file. MOC and PSSR procedures can help to manage these deficiencies.

## 8.5 EQUIPMENT DEFICIENCY COMMUNICATION

Communicating equipment deficiencies to affected personnel is critical to help ensure that incidents do not occur. Affected personnel almost always includes operators, but maintenance and contract personnel may also need to be aware of some equipment deficiencies. Communication needs to occur at three points in the equipment deficiency cycle:

- Immediate hazard and initial response. For equipment deficiencies that are discovered via equipment failures (e.g., pump seal failure), personnel discovering the deficiency must know to communicate the immediate hazard (e.g., fire, release of toxic materials) to potentially affected personnel (e.g., personnel in the area) and the actions needed to correct/mitigate the hazard (e.g., sound the evacuation alarm, stop the pump).
- Status of deficient equipment. For equipment that is shut down and for equipment operating in a deficient state (with or without temporary repairs), updated equipment status should be promptly communicated to personnel. In addition, the precautions (if any) associated with operating the deficient equipment (e.g., process running at lower pressures due to a vessel rerating) must also be clearly communicated.
- Return of deficient equipment to normal service. When the equipment deficiency is permanently corrected, personnel must be informed when the equipment is returned to service.

Communication of the immediate hazard and the initial response is typically addressed by a facility's emergency response procedures. The status of deficient equipment and return of deficient equipment to normal service can be addressed via equipment deficiency logs and/or the site's MOC program.

## 8.6 PERMANENT CORRECTION OF EQUIPMENT DEFICIENCIES

Sometimes, facilities have found that temporary repairs implemented in response to equipment deficiencies have a tendency to become "permanent." This has led to catastrophic incidents because a temporary repair, by definition, is not fit for permanent service. The best strategy to counter this tendency is to develop a system for tracking temporary repairs of deficient equipment. Typically, the equipment deficiency process uses the following systems to track temporary repairs:

- MOC procedures for temporary changes
- Equipment deficiency logs/forms
- Equipment deficiency flags/designators in the facility's computerized maintenance management system (CMMS)

For batch operations, temporary repairs often can be corrected at the end of a batch. However, correcting temporary repairs on continuous operations can be more problematic. Therefore, the equipment deficiency resolution process for these operations should dictate that temporary repairs be scheduled for termination as soon as possible and no longer than the next shutdown/turnaround. In addition, the system should allow personnel to readily identify temporary repairs so that they can be corrected in the event of an unscheduled shutdown.

## 8.7 DEFICIENCY MANAGEMENT ROLES AND RESPONSIBILITIES

The roles and responsibilities for deficiency management can be assigned to personnel in various departments. A high-level procedure that describes the management system and approval authorities used to address equipment deficiencies is recommended.

Typically, personnel in the Maintenance, Engineering, and Operations departments will be involved in managing deficiencies. Outside personnel, such as contractors, may also be involved. Example roles and responsibilities for managing deficiencies are provided in Table 8-3. The matrix designates assignments to personnel with "R" as the person(s) responsible for the activity, "A" as the approver of the work or decisions made by the responsible party, "S" as persons supporting the responsible party in completing the activity, and "I" as persons informed when the activity is completed or delayed.

## 8.8 REFERENCE

8-1. American Petroleum Institute, *Recommended Practice for Fitness-For-Service,* API RP 579, Washington, DC, 2000.

# 8. EQUIPMENT DEFICIENCY MANAGEMENT 129

**TABLE 8-3**
**Example Roles and Responsibilities Matrix for Equipment Deficiency Resolution**

| Activity | Inspection and Maintenance Personnel | | | | | | Engineering Personnel | | | Operations Personnel | | | Other Personnel | | | | |
|---|---|---|---|---|---|---|---|---|---|---|---|---|---|---|---|---|---|
| | Maintenance Manager | Maintenance Engineers | Maintenance Supervisors | Inspectors | Maintenance Technicians | Maintenance Planner/Scheduler | Engineering Manager | Project Engineers | Process Engineers | Unit Superintendent | Production Supervisors | Operators | Plant Manager | EHS Manager | Process Safety Coordinator | Contractor | Equipment Vendor |
| **Identifying Acceptance Criteria** | | | | | | | | | | | | | | | | | |
| • New equipment | | S | | S | S | | A | R | | | | | — | — | — | | S |
| • Operating equipment | | R | | S | | | A | S | S | | — | | — | — | — | S | |
| **Identifying Deficient Equipment** | | | | | | | | | | | | | | | | | |
| • New equipment | S | S | | S | S | S | S | R | | | | | — | | — | | |
| • Operating equipment | | R | | S | S | | A | | S | | A | S | — | | | | |
| **Responding to Equipment Deficiencies** | | | | | | | | | | | | | | | | | |
| • New equipment | | | | S | | | A | R | S | | | | — | — | — | | S |
| • Operating equipment | S | R | | S | | | A | | S | S | A | S | | | — | S | |

## TABLE 8-3 *(Continued)*

| Activity | Inspection and Maintenance Personnel | | | | | | Engineering Personnel | | | Operations Personnel | | | Other Personnel | | | | |
|---|---|---|---|---|---|---|---|---|---|---|---|---|---|---|---|---|---|
| | Maintenance Manager | Maintenance Engineers | Maintenance Supervisors | Inspectors | Maintenance Technicians | Maintenance Planner/Scheduler | Engineering Manager | Project Engineers | Process Engineers | Unit Superintendent | Production Supervisors | Operators | Plant Manager | EHS Manager | Process Safety Coordinator | Contractor | Equipment Vendor |
| **Tracking Equipment Deficiencies to Completion** | | | | | | | | | | | | | | | | | |
| • New equipment | | R | | S | S | | S | R | S | | | S | — | — | — | | |
| • Operating equipment | | | S | S | S | | S | R | | | R | S | — | — | — | | |
| **Equipment Deficiency Communication** | | | | | | | | | | | | | | | | | |
| • New equipment | | | | | | — | | | | | | | | | | | |
| • Operating equipment | R | S | S | | | | | | | R | S | | — | — | — | | |

## 8. EQUIPMENT DEFICIENCY MANAGEMENT

**Appendix 8A. Fitness for Service (FFS)**

An FFS assessment is performed on pressure-retaining equipment with a deficiency to justify continued service or to identify the parameters required to enter or return the equipment to service. (Note: This appendix provides a brief overview of API RP 579. Readers are encouraged to refer to API RP 579 when researching and performing FFS assessments.) The FFS assessment of pressure-retaining equipment typically involves the resolution of a deficiency found during an inspection. An FFS assessment is also used to define inspection acceptance criteria for a degraded condition, such as for large areas of metal depletion or flaws in the pressure-retaining component's metal or welding. FFS assessments can also provide alternate documentation for the suitability of equipment for service if the equipment's original records (e.g., the manufacturer's data report, Form U-1A) are not available.

A deficiency occurs when the equipment condition is outside the established limit that define equipment integrity (acceptance criteria). Typically, personnel identify deficiencies in pressure equipment during inspection or repair activities. However, inspections of new equipment fabrication and installation can also uncover deficiencies, such as gouges or dents, that may be difficult to resolve using the equipment design and fabrication codes. Such deficiencies may also be problematic because of delivery schedule and business interruption concerns. To assist facilities in evaluating these areas of concern, API RP 579 was developed to provide standardized technical evaluation methods for a wide variety of commonly identified modes of degradation in pressure-retaining equipment. API RP 579 defines FFS assessments as "engineering evaluations that are performed to demonstrate the structural integrity of an in-service component containing a flaw or damage." The FFS evaluation methodologies provide consistent engineering tools that may be applied with increasing technical rigor as necessary, based on the extent and magnitude of the equipment's degraded condition.

***Scope and Purpose of FFS.*** Facilities inspect in-service pressure-retaining equipment to characterize the condition of the equipment, which enables personnel to predict with confidence the remaining service life of the equipment. A routine inspection of pressure equipment is designed to find indications of degradation that warrant follow-up investigation, not to locate the areas of maximum degradation. Following the inspection(s), conservative evaluation criteria are initially applied to the results of the inspection to determine suitability for service. However, the evaluation of the areas of concern may require additional analysis beyond the conservative evaluation criteria derived from fabrication and inspection codes.

FFS evaluations using API RP 579 are applicable to pressure boundary components of metallic pressure vessels, piping, and storage tanks that have been designed and fabricated to a nationally recognized code or standard (e.g., American Society of Mechanical Engineers [ASME] BPV code, ANSI/ASME B31.3, API 620, API 650). FFS evaluations are to be performed under the supervision of an engineer familiar with the requirements of the applicable code.

The FFS assessment may result in a variety of outcomes resulting in follow-up actions ranging from continued service within existing safe operating limits to immediate removal from service for remediation or replacement. Common results from an FFS assessment include:

- Operating with reduced operating pressure or temperature rating
- Reducing maximum fill heights for tanks
- Changing service conditions or providing inherent process safeguards to prevent further damage
- Intensifying and/or increasing the frequency of on-stream condition monitoring (CM) to determine (1) if the degradation mode is active and (2) the rate of degradation
- Reducing service life or providing a time-specific remediation schedule
- Defining limited repair scopes through the inspection codes

In addition, an FFS assessment provides the owner-user of the pressure-retaining equipment with valuable information and options for planning and scheduling inspections and maintenance work on the equipment. This information includes:

- Equipment design basis documentation, including documents that are used to comply with regulatory requirements by establishing that equipment is safe for continued operation
- Acceptance criteria for future inspections
- Justification for the deferral of equipment replacement or repair
- Options for rerating, altering, repairing, or replacing equipment
- Forecast of remaining useful life of equipment
- Future inspection requirements

***FFS Evaluation Methodology.*** The FFS methodology, as presented in API RP 579, discusses nine equipment damage or flaw conditions for applicable pressure-retaining equipment, pressure vessels, piping, and storage tank shells. The damage or flaw conditions discussed in API RP 579 are:

- Brittle fracture
- General metal loss
- Local metal loss
- Pitting corrosion
- Blisters and laminations
- Weld misalignment and shell distortions
- Crack-like flaws
- Equipment operating in the creep range
- Fire damage

## 8. EQUIPMENT DEFICIENCY MANAGEMENT

For each of these damage and flaw conditions, the assessment techniques and the technical acceptance requirements for three levels of FFS assessments are described. Each level of assessment increases in rigor and complexity. The increases in rigor are associated with (1) the amount and detail of additional inspection data and (2) the need for specialized expertise. The three levels of assessment are designed to reach, if possible, acceptable outcomes with minimum analysis effort. The level of effort required to complete an FFS assessment is consistent with the severity level of the degradation, the economics of the equipment's remediation, and the economics of the necessary business interruption. Typically, engineers or inspectors who are familiar with the equipment fabrication codes can perform a Level 1 FFS assessment. Levels 2 and 3 require individuals with specialized knowledge in fracture mechanics, materials science, nondestructive testing (NDT), and structural engineering. A Level 2 analysis requires information similar to that required for Level 1, but uses more detailed calculations. A Level 3 analysis is generally a numerical finite element model requiring the involvement of personnel knowledgeable in Level 2 evaluations, but who are also skilled in the setup, running, and interpretation of the numerical model's results.

For each damage or flaw condition, a damage-specific procedure for conducting an FFS assessment is discussed. API RP 579 describes an eight-step procedure for conducting an FFS assessment for each damage condition of a component:

1. Flaw and damage mechanism identification
2. Applicability and limitations of the FFS assessment procedures
3. Data requirements for conducting the FFS assessment
4. FFS assessment techniques and acceptance criteria
5. Remaining life evaluation
6. Equipment or component remediation
7. On-stream monitoring of the component or equipment
8. Documentation of the FFS assessment

Flaw and damage mechanism identification is not specifically covered in API RP 579. This step involves understanding the results of the applied inspection technique in order to properly characterize the damage condition(s) to be evaluated during the FFS assessment. API RP 579 provides a list of RAGAGEPs that are useful for damage and flaw mechanism identification and characterization. (In addition, API RP 571, *Damage Mechanisms Affecting Fixed Equipment in the Refining Industry*, provides additional information on fixed equipment damage mechanisms.)

API RP 579 provides specific guidance on Steps 2 through 8 for each flaw and damage mechanism identified. The data requirements and documentation requirements for these steps are similar to those required for typical in-service inspection plans.

# 9
# EQUIPMENT-SPECIFIC INTEGRITY MANAGEMENT

The previous chapters in this book focused on the programmatic aspects (e.g., management activities and systems) of a facility's mechanical integrity (MI) program. This chapter provides detailed MI program information for common types of equipment. The information provides facility personnel with guidance and suggestions for managing the integrity of particular equipment items included in the MI program. Specifically, the information included in this chapter can assist facilities in selecting inspection, testing, and preventive maintenance (ITPM) tasks (see Chapter 4), identifying appropriate quality assurance (QA) activities (see Chapter 7), and initiating appropriate repairs (i.e., corrections for deficient conditions referenced in Chapter 8).

Much of the information in this chapter is documented in a series of matrices for different types of equipment that might require equipment-specific procedures, training, and/or tasks. Each matrix is organized into the four phases of equipment integrity introduced in Chapter 1. The objective of each phase is explained below:

1. New equipment design, fabrication, and installation. During this phase, activities focus on ensuring that new equipment is suitable for its intended service; therefore, many of the activities in this phase are directly related to the QA activities for the early part of the equipment life cycle discussed in Chapter 7.
2. Inspection and testing. During this phase, activities focus on ensuring the ongoing integrity of equipment or functionality of equipment safeguards for a specified inspection and testing interval. These activities are included in the ITPM plan discussed in Chapter 4.
3. Preventive maintenance. During this phase, activities focus on preventing premature failure of the equipment and its components, and can include performing servicing tasks (e.g., lubrication) and/or inspecting and replacing components that are subject to wear. These activities also are part of the ITPM plan discussed in Chapter 4.

4. Repair. During this phase, activities focus on responding to equipment failures, and repairing and returning equipment to service in a condition suitable for its intended use. Such repair activities are similar to the practices discussed in Chapter 8 regarding correction of equipment deficiencies.

In the matrices, these four phases are in separate columns, and specific information for each phase is provided in the rows. The information has been divided into the following categories:

1. Example activities and typical frequencies. Actions listed for the equipment design, fabrication, and installation phase include QA activities for that type of equipment, as well as the timing of these activities during the design, fabrication, and installation cycle. For the inspection and testing and preventive maintenance (PM) phases, the activities and frequencies provide the typical information needed for the ITPM plan for that type of equipment. The activities listed for the repair phase include common repairs as well as some repair QA activities.
2. Technical basis for activity and frequency. This row provides the jurisdictional requirement, code or standard, or other rationale (e.g., common industry practice, manufacturers' recommendation) typically used to justify the activity and frequency.
3. Sources of acceptance criteria. Information sources that typically contain acceptance criteria for MI activities are provided in this row.
4. Typical failures of interest. This row lists equipment failures that can affect equipment integrity.
5. Personnel qualifications. This row provides general and specific qualification requirements for personnel performing the activities.
6. Procedure requirements. The information in this row outlines types of procedures typically developed for the phase and/or the general content of the procedures.
7. Documentation requirements. This row lists typical documentation created from the activities as well as documentation retention recommendations. In addition, suggestions for documenting activities by exception are provided.

The following sections of this chapter each contain a representative matrix for these different classes of equipment:

- Fixed equipment
- Relief and vent systems
- Instrumentation and controls
- Rotating equipment
- Fired equipment
- Electrical systems

## 9. EQUIPMENT-SPECIFIC INTEGRITY MANAGEMENT

In addition, each section also provides (1) a list of the specific types of equipment within the equipment class that have matrices on the CD accompanying this book, (2) information about relevant recognized and generally accepted good engineering practices (RAGAGEPs), and (3) discussion items (i.e., general information for the equipment class or detailed information beyond what is included in the matrix).

Two sections included in this chapter, Fire Protection Systems and Miscellaneous Equipment, present relevant information specific to the equipment; however, no matrices are included. Matrices have not been included for fire protection systems because volumes of published information are generally already accessible to most facility personnel. In addition, the specific ITPM activities for fire protection systems are usually provided by the authority having jurisdiction (e.g., local fire marshal, company insurance carrier). Second, specific matrices for miscellaneous systems have not been included in the book because of the wide variety of designs and issues with these systems. However, information on applicable RAGAGEPs and a brief discussion of MI activities are provided.

## 9.1 FIXED EQUIPMENT

The design, inspection, and testing requirements for most types of fixed equipment are governed by a RAGAGEP. The primary organizations issuing RAGAGEPs for fixed equipment include the American Society of Mechanical Engineers (ASME), the American Petroleum Institute (API), and the National Board of Boiler and Pressure Vessel Inspectors (NBBPVI). Tables 9-1, 9-2, and 9-3 provide brief summaries of a few of the more common RAGAGEPs for pressure vessels, atmospheric and low-pressure storage tanks, and process piping, respectively. In addition, other RAGAGEPs have been developed for certain applications and specific chemicals. These include American National Standards Institute (ANSI)-K61.1, *Safety Requirements for the Storage and Handling of Anhydrous Ammonia* for processes using anhydrous ammonia, as well as the codes and standards developed by the Chlorine Institute (CI) for processes using chlorine and the International Institute of Ammonia Refrigeration (IIAR) for ammonia refrigeration installations.

## TABLE 9-1
## RAGAGEPs for Pressure Vessels

| Issuing Organization | Document | | Application |
|---|---|---|---|
| | Number | Title | |
| API | API 510 | Pressure Vessel Inspection Code: Maintenance Inspection, Rating, Repair, and Alteration | Covers the maintenance, inspection, repair, alteration, and rerating procedures for pressure vessels. |
| API | Recommended Practice (RP) 572 | Inspection of Pressure Vessels | Covers the inspection of pressure vessels. |
| API | ANSI/API 660 or International Organization for Standardization (ISO) 16812 | Shell-and-Tube Heat Exchangers for General Refinery Services | Defines the minimum requirements for the mechanical design, material selection, fabrication, inspection, testing, and preparation for shipment of shell-and-tube heat exchangers. |
| ASME | ASME Code, Section VIII | ASME Boiler and Pressure Vessel Code (BPVC), Unfired Pressure Vessels | Provides requirements applicable to the design, fabrication, inspection, testing, and certification of pressure vessels operating at either internal or external pressures exceeding 15 psig. |
| NBBPVI | National Board (NB)-23 | National Board Inspection Code | Provides rules and guidelines for in-service inspection of boilers, pressure vessels, piping, and pressure relief valves (PRVs). Also provides rules for the repair, alteration, and rerating of pressure-retaining items and for the repair of PRVs. |

## 9. EQUIPMENT-SPECIFIC INTEGRITY MANAGEMENT

### TABLE 9-2
### RAGAGEPs for Atmospheric and Low-pressure Storage Tanks

| Issuing Organization | Document Number | Document Title | Application |
|---|---|---|---|
| API | RP 575 | Inspection of Atmospheric and Low Pressure Storage Tanks | Covers the inspection of atmospheric and low-pressure storage tanks designed to operate at pressures from atmospheric to 15 psig. |
| API | 620 | Design and Construction of Large, Welded, Low-pressure Storage Tanks | Covers the design and construction of large, welded, low-pressure carbon steel aboveground storage tanks, including flat-bottom tanks that have a single vertical axis of revolution. Applies to tanks with pressure in their vapor spaces at not more than 15 psig. |
| API | 650 | Welded Steel Tanks for Oil Storage | Covers the material, fabrication, erection, and testing requirements for aboveground, vertical, cylindrical, closed- and open-top, welded steel storage tanks. Applies to tanks with internal pressures approximating atmospheric pressure. |
| API | 653 | Tank Inspection, Repair, Alteration, and Reconstruction | Covers the inspection, repair, alteration, and reconstruction of steel aboveground storage tanks. |

## TABLE 9-3
### RAGAGEPs for Process Piping

| Issuing Organization | Document | | Application |
| --- | --- | --- | --- |
| | Number | Title | |
| ANSI | ANSI/ASME B31.3 | B31.3 - Process Piping | Provides requirements for materials and components, design, fabrication, assembly, erection, examination, inspection, and testing of piping. |
| API | API 570 | Piping Inspection Code: Inspection, Repair, Alteration, and Rerating of In-Service Piping Systems | Provides procedures for the inspection, repair, alteration, and rerating of metallic piping that have been in-service. |
| API | API RP 574 | Inspection Practices for Piping System Components | Covers inspection practices for piping, tubing, valves (not including control valves), and fittings. This document is a supplement to ANSI/API 570. |
| NBBPVI | NB-23 | National Board Inspection Code | Provides rules and guidelines for in-service inspection of boilers, pressure vessels, piping, and PRVs. Also provides rules for the repair, alteration, and rerating of pressure-retaining items and for the repair of PRVs. |

## 9. EQUIPMENT-SPECIFIC INTEGRITY MANAGEMENT

In general, the RAGAGEPs contain design, inspection, and testing requirements needed to ensure structural integrity of the fixed equipment items. Many times, however, these documents do not include process performance requirements that might be important for ensuring process safety. Therefore, additional design considerations, inspections, and tests may be needed to ensure process safety. Common examples include:

- Performance testing and monitoring of scrubber operations
- Performance testing and monitoring of heat exchange equipment when failure to adequately remove heat can have process safety implications (e.g., removal of heat from an exothermic reaction)
- Lubrication of swivel joints in loading arms to ensure that the joint is sealed

Design, fabrication, and testing requirements for fiberglass vessels, tanks, and equipment are contained in several RAGAGEPs published by API, ASME, and the American Society of Testing and Materials (ASTM) International. However, these RAGAGEPs generally do not provide specific guidance on the inspection and testing of in-service equipment. Common RAGAGEPs for fiberglass constructed equipment are:

- ASME Section X — Fiberglass-Reinforced Plastic Pressure Vessels
- API Spec 12P.— Specification for Fiberglass-Reinforced Plastic Tanks
- ASTM D3299 — Filament-Wound Fiberglass-Reinforced Plastic Chemical Resistant Tanks
- ASTM D2563 — Recommended Practice for Classifying Visual Defects in Laminates
- ASTM D2583 — Indentation Hardness of Rigid Plastics Using a Barcol Impressor

Transportation equipment that remains stationary at a facility (e.g., tank trucks, railcars) should be included in the MI program as a special category of "fixed" equipment. In the United States, the design, inspection, and testing requirements for this equipment are defined by Department of Transportation (DOT) regulations (*49 Code of Federal Regulations [CFR]* Chapter 1, Subchapter C). For most organizations, the company shipping the materials and/or the owner of the tank truck/railcar is responsible for ensuring the integrity of this equipment. Therefore, most facilities will need to ensure only that the equipment is in compliance with the DOT regulations.

Tables 9-13 and 9-14 (in Section 9.9) depict MI activity matrices for pressure vessels (including columns, filters, and heat exchangers) and piping systems, respectively. The CD accompanying this book includes these matrices as well as a matrix for storage tanks.

## 9.2 RELIEF AND VENT SYSTEMS

The design, inspection, and testing requirements for most types of pressure relief equipment are governed by a RAGAGEP. These requirements are often included in the RAGAGEP for the fixed equipment on which the relief device is installed (e.g., pressure vessel, atmospheric storage tank). ASME, API, NBBPVI, and the National Fire Protection Association (NFPA) are the primary organizations issuing RAGAGEPs for these pressure relief devices. Brief summaries of some common RAGAGEPs for pressure relieving devices (i.e., pressure relief valves [PRVs], rupture disks, and vacuum PRVs) are presented in Table 9-4. In addition, the Center for Chemical Process Safety (CCPS) book, *Emergency Relief System Design Using DIERs Technology*, provides guidance on sizing relief systems involving reactive materials and two-phase flow situations.

Also, other RAGAGEPs have been developed for certain applications and select chemicals. These include (1) ANSI-K61.1 for processes using anhydrous ammonia, (2) codes and standards developed by the Chlorine Institute (CI) for processes using chlorine, (3) IIAR standards for ammonia refrigeration installations, and (4) standards developed by the Compressed Gas Association and Underwriters Laboratories Inc. (UL) for selected applications. Several NPFA standards and Occupational Safety and Health Administration (OSHA) regulations provide guidance on pressure relief device design and maintenance; however, these standards often reference requirements in other RAGAGEPs.

RAGAGEPs for other relief system equipment (e.g., flame/detonation arresters, emergency vents, vent headers, thermal oxidizers) are not as widely published. Many times, the design criteria are based on the process and on manufacturer-supplied design criteria. In addition, ITPM activities for this equipment are typically based on a combination of manufacturers' recommendations and operating experience and knowledge.

Factory Mutual Research (FM), UL, and ASTM International have established standards/guidelines for testing and approving flame/detonation arresters. In addition, the United States Coast Guard (USCG) developed regulations for marine vapor control that (1) provide requirements for the installation of a flame detonation arrester for marine applications and (2) contain guidelines for testing and approving arresters. The CCPS book, *Deflagration and Detonation Arresters*, also provides information on designing, installing, inspecting, and maintaining arresters.

Piping associated with relief and vent systems (e.g., vent piping, relief inlet and discharge piping, thermal oxidizer piping) is typically designed, fabricated, and installed in accordance with the chemical process piping standard, ANSI/ASME B31.3. Special design issues, such as allowable pressure drop and support requirements, and relief piping concerns, such as plugging, are identified in applicable relief device RAGAGEPs (e.g., API Recommended Practice [RP] 572, Part II). Relief and vent system piping should be inspected in accordance with the piping inspection code, API 570.

9. EQUIPMENT-SPECIFIC INTEGRITY MANAGEMENT 143

TABLE 9-4
RAGAGEPs for Pressure Relieving Devices

| Issuing Organization | Document | | Application |
|---|---|---|---|
| | Number | Title | |
| ANSI/ASME | ANSI/ASME B31.3 | B31.3 - Process Piping | Provides requirements for materials and components, design, fabrication, assembly, erection, examination, inspection, and testing of piping. |
| API | API 510 | Pressure Vessel Inspection Code: Maintenance, Inspection, Rating, Repair, and Alteration | Covers the maintenance, inspection, repair, alteration, and rerating procedures for pressure vessels used by the petroleum and chemical process industries. Also covers inspection and testing of pressure relief devices. |
| API | RP 520 | Sizing, Selection, and Installation of Pressure-relieving Devices in Refineries, Part I - Sizing and Selection, Part II - Installation | Covers the sizing, selection, and installation of pressure relief devices for equipment that has a maximum allowable working pressure of 15 psig or greater. |
| API | RP 576 | Inspection of Pressure Relieving Devices | Describes the inspection and repair practices for automatic pressure-relieving devices, including pressure safety valves (PSVs), pilot-operated PRVs, rupture disks, and weight-loaded pressure vacuum vents. |
| ASME | ASME B&PV Code, Section I | ASME BPVC - Power Boilers | Provides requirements for all methods of construction and relief protection of power, electric, and miniature boilers, as well as high-temperature water boilers used in stationary service. |
| ASME | ASME B&PV Code, Section IV | ASME BPVC - Heating Boilers | Provides requirements for design, fabrication, installation, and inspection of steam generating boilers, and hot water boilers intended for low-pressure service that are directly fired by oil, gas, electricity, or coal. It also covers methods of checking safety valve and safety relief valve capacity. |
| ASME | ASME B&PV Code, Section VIII | ASME BPVC - Pressure Vessels | Provides requirements applicable to the design, fabrication, inspection, testing, and certification of pressure vessels operating at either internal or external pressures exceeding 15 psig. Also provides requirements for pressure vessel relief devices. |
| NBBPVI | NB-23 | National Board Inspection Code | Provides rules and guidelines for in-service inspection of boilers, pressure vessels, piping, and PSVs. Also provides rules for the repair, alteration, and rerating of pressure-retaining items and for the repair of PSVs. |

Table 9-15 (in Section 9.9) presents an MI activity matrix for PRVs. The CD accompanying this book also includes this matrix, as well as additional matrices for the following types of relief and vent systems:

- Rupture disks
- Vacuum PRVs/conservation vents (i.e., low pressure/vacuum relief devices)
- Flame/detonation arresters
- Emergency vents
- Vent headers
- Thermal oxidizers
- Flares

## 9.3 INSTRUMENTATION AND CONTROLS

RAGAGEPs published by API and the Instrumentation, Systems, and Automation Society (ISA) govern the design, inspection, and testing requirements for some instrumentation. In addition, ASME and NFPA have published codes that are applicable to burner management systems. Table 9-5 provides brief summaries of common RAGAGEPs for instrumentation and controls. Other RAGAGEPs, such as CI and IIAR publications, and ANSI-K61.1, also contain information on instrumentation and controls. Specifically, CI has a publication outlining requirements for chlorine detection systems (Reference 9-1), and IIAR standards and ANSI–K61.1 provide information on instrumentation and controls for ammonia refrigeration and anhydrous ammonia installations, respectively.

In addition to the RAGAGEPs listed in Table 9-5, ISA has issued several other publications that provide information on the design and maintenance of instrumentation and control systems. The CCPS book, *Guidelines for Safe Automation of Chemical Processes*, also contains information on designing and maintaining safety instrumented systems (SISs).

Table 9-16 (in Section 9.9) is an MI activity matrix for SISs and emergency shutdowns (ESDs). The CD accompanying this book also includes this matrix, as well as additional matrices for the following types of instrumentation and controls:

- Critical process controls
- Critical alarms and interlocks
- Toxic chemical monitors and detection systems
- Flammable area monitors and detection systems
- Conductivity, pH, and other process analyzers
- Burner management systems

## 9. EQUIPMENT-SPECIFIC INTEGRITY MANAGEMENT

## 9.4 ROTATING EQUIPMENT

Most of the RAGAGEPs available for rotating equipment cover the design, fabrication, and installation of the equipment; very few RAGAGEPs are available for ITPM activities. Therefore, many of the inspection and testing requirements are derived from the manufacturers' recommendations, common industry practices (e.g., vibration analysis), and operating experience and history. In addition, some organizations use analysis techniques, such as reliability-centered maintenance (RCM), to determine ITPM activities for rotating equipment.

ANSI, API, and the Hydraulic Institute (HI) publish RAGAGEPs for pumps, compressors, fans, turbines, and gearboxes. Brief summaries of some common RAGAGEPs from ANSI and API for pumps, compressors, turbines, and fans and gearboxes are presented in Tables 9-6 through 9-9, respectively.

Other RAGAGEPs available include standards and bulletins published by IIAR that contain specific information on the design, installation, inspection, and maintenance of ammonia refrigeration equipment. Also, design, inspection, and test requirements for electric motors are provided in NPFA 70, *National Electric Code*, and NFPA 70B, *Recommended Practice for Electrical Equipment Maintenance*.

Table 9-17 (in Section 9.9) provides an MI activity matrix for pumps. The CD accompanying this book also includes this matrix, as well as additional matrices for the following types of rotating equipment:

- Reciprocating compressors
- Centrifugal compressors, including specific protection systems (e.g., pressure cutouts)
- Process fans and blowers
- Agitators and mixers
- Electric motors
- Gas turbines
- Steam turbines
- Gearboxes

## TABLE 9-5
## RAGAGEPs for Instrumentation and Controls

| Issuing Organization | Document Number | Document Title | Application |
|---|---|---|---|
| API | RP 551 | Process Measurement Instrumentation | Provides procedures for installation of the more generally used measuring and control instruments and related accessories. |
| API | RP 554 | Process Instrumentation and Control | Covers performance requirements and considerations for the selection, specification, installation, and testing of process instrumentation and control systems. |
| API | API 555 | Process Analyzers | Addresses the associated systems, installation, and maintenance of analyzers. |
| ASME | CSD-1 | Controls and Safety Devices for Automatically Fired Boilers | Covers requirements for the assembly, maintenance, and operation of controls and safety devices installed on automatically operated boilers that are directly fired with gas, oil, gas-oil, or electricity, subject to certain service limitations and exclusions. |
| International Electrotechnical Commission (IEC) | IEC 61508-SER | Functional safety of electrical/electronic/programmable electronic safety-related systems | Sets out a generic approach for all safety life-cycle activities for systems comprising electrical, electronic, and/or programmable electronic components that are used to perform safety functions. |
| ISA | ISA S84.01 | Application of SISs for the Process Industries | Provides a safety life cycle and requirements for design, installation, and maintenance of SISs. |
| ISA | ISA-TR-84.00.02 Parts 1 through 5 | Safety Instrumented Functions (SIFs) – Safety Integrity Level (SIL) Evaluation Techniques | This series covers the different evaluation techniques that can be used to determine if a specific SIS design satisfies the SIL requirements defined in the SIF. |
| ISA | ISA-91.00.01 | Identification of Emergency Shutdown (ESD) Systems and Controls that Are Critical to Maintaining Safety in the Process Industries | Provides general requirements for determining safety-critical ESDs and controls, and for maintaining the identified instrumentation. |
| NFPA | NFPA 85 | Boiler and Combustion Systems Hazards Code | Discusses the fundamentals, maintenance, inspection, training, and safety for the reduction of combustion system hazards. |

## 9. EQUIPMENT-SPECIFIC INTEGRITY MANAGEMENT

TABLE 9-6
RAGAGEPs for Pumps

| Issuing Organization | Document Number | Document Title | Application |
|---|---|---|---|
| ANSI | ANSI/ASME B73.1 | Specification for Horizontal End Suction Centrifugal Pumps for Chemical Process | Covers centrifugal pumps of horizontal, end suction single stage, centerline discharge design. Includes dimensional interchangeability requirements and certain design features to facilitate installation and maintenance. |
| ANSI | ANSI/ASME B73.2 | Specification for Vertical In-Line Centrifugal Pumps for Chemical Process | Covers motor-driven centrifugal pumps of vertical shaft, single stage design with suction and discharge nozzles in-line. Includes dimensional interchangeability requirements and certain design features to facilitate installation and maintenance. |
| ANSI | ANSI/ASME B73.3 | Specification for Sealless Horizontal End Suction Centrifugal Pumps for Chemical Process | Covers sealless centrifugal pumps of horizontal, end suction single stage, centerline discharge design. Includes dimensional interchangeability and features to facilitate installation and maintenance. |
| API | 610 | Centrifugal Pumps for Petroleum, Petrochemical, and Natural Gas Industries | Specifies requirements for centrifugal pumps, including running in reverse as hydraulic power turbines. |
| API | 674 | Positive Displacement Pumps - Reciprocating | Covers the minimum requirements for reciprocating positive displacement pumps. |
| API | 675 | Positive Displacement Pumps - Controlled Volume | Covers the minimum requirements for controlled volume positive displacement pumps. |
| API | 676 | Positive Displacement Pumps - Rotary | Covers the minimum requirements for rotary positive displacement pumps. |
| API | 681 | Liquid Ring Vacuum Pumps and Compressors | Defines the minimum requirements for the basic design, inspection, testing, and preparation for shipment of liquid ring vacuum pumps and compressors. |
| API | 682 | Pumps - Shaft Sealing Systems for Centrifugal and Rotary Pumps | Specifies requirements and provides suggestions for sealing systems for centrifugal and rotary pumps. |

## TABLE 9-7
## RAGAGEPs for Compressors

| Issuing Organization | Document | | Application |
|---|---|---|---|
| | Number | Title | |
| ANSI | ANSI/ASME B19.3 | Safety Standard for Compressors for Process Industries | Covers the requirements for safety devices and protective facilities to help prevent compressor accidents as a result of excessive pressure, destructive mechanical failures, internal fires or explosions, and leakage of toxic or flammable fluids. |
| API | 617 | Axial and Centrifugal Compressors and Expander-compressors for Petroleum, Chemical, and Gas Industry Services | Covers the minimum requirements for centrifugal compressors that handle air or gas. |
| API | 618 | Reciprocating Compressors for Petroleum, Chemical, and Gas Industry Services | Covers the minimum requirements for reciprocating compressors and their drivers handling process air or gas with either lubricated or nonlubricated cylinders. |
| API | 681 | Liquid Ring Vacuum Pumps and Compressors | Defines the minimum requirements for the basic design, inspection, testing, and preparation for shipment of liquid ring vacuum pumps and compressors. |

## 9. EQUIPMENT-SPECIFIC INTEGRITY MANAGEMENT

**TABLE 9-8**
**RAGAGEPs for Turbines**

| Issuing Organization | Document Number | Document Title | Application |
|---|---|---|---|
| API | 616 | Gas Turbines for the Petroleum, Chemical, and Gas Industry Services | Covers the minimum requirements for open, simple, and regenerative-cycle combustion gas turbine units for services of mechanical drive, generator drive, or process gas generation. |
| API | 611 | General Purpose Steam Turbines for Petroleum, Chemical, and Gas Industry Service | Covers the minimum requirements for the basic design, materials, related lubrication systems, controls, auxiliary equipment, and accessories for general-purpose steam turbines. |
| API | 612 | Petroleum, Petrochemical and Natural Gas Industries - Steam Turbines - Special-purpose Applications | Specifies requirements and gives recommendations for the design, materials, fabrication, inspection, testing, and preparation for shipment of special-service steam turbines. |

## TABLE 9-9
## RAGAGEPs for Fans and Gearboxes

| Issuing Organization | Document | | Application |
| --- | --- | --- | --- |
| | Number | Title | |
| API | 673 | Special Purpose Fans | Covers the minimum requirements for centrifugal fans intended for continuous duty. |
| API | 613 | Special Purpose Gear Units for Petroleum, Chemical, and Gas Industry Services | Covers the minimum requirements for special-purpose, enclosed, precision, single- and double-helical, one- and two-stage speed increasers and reducers of parallel-shaft design. |
| API | 677 | General-purpose Gear Units for Petroleum, Chemical, and Gas Industry Services | Covers the minimum requirements for general-purpose, enclosed, single- and multi-stage gear units incorporating parallel-shaft helical and right-angle bevel gears. |

## 9. EQUIPMENT-SPECIFIC INTEGRITY MANAGEMENT

## 9.5 FIRED EQUIPMENT

The applicable RAGAGEPs for fired equipment are grouped based on application (1) boilers and (2) fired heaters and furnaces. Boiler design requirements are governed by Sections I and IV of the ASME Boiler and Pressure Vessel Code (BPVC), and inspection and testing requirements are governed by NBBPVI's NB-23. API publishes most of the RAGAGEPs for fired heaters and furnaces in process industry service. Table 9-10 lists the common RAGAGEPs for fired heaters and furnaces.

NFPA and some industrial insurers have also published documents that provide information on the design, installation, and maintenance of fired heaters and furnaces (References 9-2 through 9-8). Specifically, NFPA has written several codes that address furnaces for special applications. The burner control system is a key element of boilers, fired heaters, and furnaces; information on these systems is provided in the instrument and control section of this chapter (see Section 9.3).

Table 9-18 (in Section 9.9) presents an MI activity matrix for fired heaters, furnaces, and boilers.

## 9.6 ELECTRICAL SYSTEMS

NFPA is a primary source of RAGAGEPs governing the design, inspection, and testing of electrical equipment. The National Electric Code (NEC) (NFPA 70) provides design information, and NFPA 70B, *Recommended Practice for Electrical Equipment Maintenance*, provides information on inspection and testing activities for most types of electrical equipment. In addition, NFPA 111, *Standard on Stored Electric Energy, Emergency, and Standby Power Systems*, provides design, inspection, and testing information for emergency generators and uninterruptible power supplies (UPSs). IEEE 446, *Recommended Practice for Emergency and Standby Power Systems for Industrial and Commercial Applications*, also provides information on the uses, power sources, design, and maintenance of emergency and standby power systems.

Table 9-19 (in Section 9.9) presents an MI activity matrix for switch gear. The CD accompanying this book also includes the matrix, as well as additional matrices for the following types of electrical equipment:

- Transformers
- Motor controls
- UPSs
- Emergency generators
- Lightning protection
- Grounding systems

TABLE 9-10
RAGAGEPs for Fired Heaters and Furnaces

| Issuing Organization | Document Number | Document Title | Application |
|---|---|---|---|
| API | 673 | Special Purpose Fans | Covers the minimum requirements for centrifugal fans intended for continuous duty. |
| API | 613 | Special Purpose Gear Units for Petroleum, Chemical, and Gas Industry Services | Covers the minimum requirements for special-purpose, enclosed, precision, single- and double-helical, one- and two-stage speed increasers and reducers of parallel-shaft design. |
| API | 677 | General-purpose Gear Units for Petroleum, Chemical, and Gas Industry Services | Covers the minimum requirements for general-purpose, enclosed, single- and multi-stage gear units incorporating parallel-shaft helical and right-angle bevel gears. |
| API | 535 | Burners for Fired Heaters in General Refinery Service | Provides guidelines for selection and/or evaluation of burners installed in fired heaters. |
| API | 560 | Fired Heaters for General Refinery Service | Covers minimum requirements for the design, materials, fabrication, inspection, testing, and preparation for shipment for fired heaters. |
| API | RP 573 | Inspection of Fired Boilers and Heaters | Covers inspection practices for fired boilers and process heaters (furnaces). |

9. EQUIPMENT-SPECIFIC INTEGRITY MANAGEMENT 153

TABLE 9-11
Summary of Commonly Used NFPA Codes for Fire Protection Systems

| Fire Protection Systems | NFPA Code | |
|---|---|---|
| | Design and Installation Requirements | Inspection, Test, and Maintenance Requirements |
| Fire detection and alarm systems | NFPA 72 | NFPA 72 |
| Automatic sprinkler systems | NFPA 13 | NFPA 25 |
| Water spray systems | NFPA 15 | NFPA 25 |
| Foam-water sprinkler systems | NFPA 16 | — |
| Foam systems | NFPA 11 | NFPA 25 |
| Standpipe and hose systems | NFPA 14 | NFPA 25 and 1962 |
| Fire pumps | NFPA 20 | NFPA 25 |
| Water supply systems | NFPA 22 and 24 | NFPA 25 |
| Fire hydrants | NFPA 24 | NFPA 25 |
| Portable fire extinguishers | NFPA 10 | NFPA 10 |
| Fire doors and dampers | NFPA 80 and 90A | —. |
| Halon systems | NFPA 12A | NFPA 12A |
| Carbon dioxide systems | NFPA 12 | NFPA 12 |
| Clean agent systems | NFPA 2001 | NFPA 2001 |
| Dry chemical extinguishing systems | NFPA 17 | NFPA 17 |
| Buildings and structures | NFPA 101 | NFPA 101 |

## 9.7 FIRE PROTECTION SYSTEMS

NFPA codes provide extensive information regarding equipment design and installation as well as ITPM information for fire protection systems. The information presented in Table 9-11 documents some of the NFPA codes that contain (1) design and installation requirements and (2) ITPM requirements.

NFPA publication, *Fire Protection Systems Inspection, Test and Maintenance Manual*, provides a thorough examination and presentation of inspection, test, and maintenance requirements. Facility personnel should consult with the authority having jurisdiction (e.g., fire marshal, insurance underwriter) when evaluating inspection, test, and maintenance requirements.

## 9.8 MISCELLANEOUS EQUIPMENT

This section discusses MI activities for the following types of equipment:

- Ventilation and purge systems
- Protective systems, such as systems for maintaining vapor space concentrations, gas fuel purge systems, exothermic reaction "kill" systems, and water curtains
- Solids-handling equipment
- Safety-critical utilities
- Other safety equipment, including eyewashes/safety showers, emergency alarms, emergency response equipment

Specifically, the following subsections identify common RAGAGEPs that provide information on design, inspection, and test requirements. General guidance for ITPM activities is also provided. In general, the training, procedure, and documentation requirements for these types of equipment are not unique; therefore, the general guidance provided in Chapters 4, 5, and 7 on these topics should be followed.

### 9.8.1 Ventilation and Purge Systems

Ventilation and purge systems are typically needed for (1) industrial hygiene purposes, (2) places of safe refuge from toxic equipment releases, and (3) electrical classification. The design, inspection, and testing requirements of ventilation and purge systems in industrial hygiene applications are dependent on many variables, such as the chemical(s) of concern, the activity being performed (e.g., sampling, connecting of railcars), and the equipment configuration. The American Conference of Governmental Industrial Hygienists (ACGIH) and other organizations have published various standards and guidelines documents, including *Design of Industrial Ventilation Systems*, that provide design requirements for various situations. These publications also contain general information on inspection and testing activities. In general, inspection and testing activities involve function and/or performance testing of the system as a whole and/or of specific system components (e.g., air handling equipment).

Ventilation and purge systems for electrical classification typically involve pressurizing a room or building (e.g., motor control center, analyzer room) or an electrical equipment enclosure (e.g., electrical panel) to prevent the ingress of flammable vapors and gases and/or combustible dusts. NFPA 496, *Standard for Purged and Pressurized Enclosures for Electrical Equipment*, provides design requirements for different (1) electrical classifications, (2) types of electrical enclosures, (3) control rooms, and (4) analyzer rooms. In addition, NFPA 70 provides additional information on the types of electrical equipment permitted and the installation of equipment in different electrical classification areas.

## 9. EQUIPMENT-SPECIFIC INTEGRITY MANAGEMENT

Ventilation for places of safe refuge typically includes (1) a pressurized room or building and (2) a means to prevent the ingress of contaminated outside air. The design codes usually do not specify ITPM requirements for building ventilation, enclosure ventilation, and purge systems. Typical ITPM activities include (1) visual inspection of the installation, (2) testing and calibration of instrumentation used to detect loss of enclosure pressure and/or presence of a flammable gas/vapor, (3) function testing of alarms, automatic power disconnects, and/or air intake dampers, and (4) PM of ventilation equipment (e.g., fans used to pressurize rooms).

### 9.8.2 Protective Systems

This section discusses common protective systems, such as systems for maintaining vapor space concentrations, gas fuel purge systems, exothermic reaction "kill" systems, and chemical water curtains, used in the chemical process industries (CPI).

Systems for maintaining vapor space concentrations are typically installed on storage tanks and/or reactors to maintain the equipment's vapor space below the lower flammable limit or above the upper flammable limit. These systems typically involve introducing an inert gas, typically nitrogen or carbon dioxide, or a flammable gas (e.g., natural gas) into the equipment vapor space. The specifics of these systems (e.g., continuous purging, batch purging, vacuum purging) determine the design and operation of the system. NFPA 69, *Standard on Explosion Prevention Systems*, describes and provides design requirements for different purging systems. However, this standard does not contain significant information on the inspection and testing requirements of these systems. Typically, the ITPM activities for these systems include visual inspection, function testing, and/or PM of critical system components (e.g., testing of pressure alarms, rebuilding of supply gas regulators). Boilers and other fired equipment must be purged of fuel gas prior to initial ignition of the burner; this activity requires a specific type of protective purge system. NFPA 54, *National Fuel Gas Code*, provides design requirements and inspection and testing guidelines for these systems.

Facilities that involve exothermic reactions often have systems installed to stop potential runaway reactions. These protective systems may (1) use a chemical injection to stop a reaction (e.g., polymerization chain inhibitor), (2) de-inventory the reactor, and/or (3) "quench" the reaction with another chemical (many times water). These automated systems typically involve the movement of material into and/or out of the reaction vessel. The design requirements for these systems are process dependent, and no consensus codes and standards exist for these systems. ITPM activities typically include a function test of the system operation and maintenance of the critical system/components as described in the applicable sections above.

While no consensus codes and standards exist for exothermic reaction safety systems, several CCPS publications provide information on the design, installation, inspection and testing, and maintenance of these systems. The information in the following CCPS publications should be considered when developing MI activities for these types of systems:

- *Essential Practices for Managing Chemical Reactivity Hazards*
- *Guidelines for Chemical Reactivity Evaluation and Application to Process Design*
- *Guidelines for Safe Storage and Handling of Reactive Materials*
- *Guidelines for Process Safety in Batch Reaction Systems*

Water curtains are installed in some facilities to knock down released chemical vapors. The design requirements for these systems are usually governed by the codes and standards for the chemical involved. Specifically, ANSI-K61.1, *Safety Requirements for the Storage and Handling of Anhydrous Ammonia*, provides design requirements for water curtains for ammonia systems. Also, API RP 751, *Safe Operation of Hydrofluoric Acid Alkylation Units*, contains design information and guidance on inspection and testing of water curtains used in HF acid alkylation units. The ITPM activities typically include a function test of the water curtain and calibration and testing of any detection and activation systems. In addition, the ITPM plan should include activities to verify the availability and integrity of any water/chemical containment system. The CCPS book, *Guidelines for Post-release Mitigation in the Chemical Process Industry*, also provides information on water curtains.

### 9.8.3 Solids-handling Systems

A primary objective of the MI program for solids-handling equipment is to prevent conditions that can result in fires and explosions caused by combustible dusts. This typically involves (1) reducing the amount of dust generated, (2) controlling the dust, and (3) preventing ignition sources. NFPA 654, *Standard for Prevention of Fire and Dust Explosion for the Manufacturing, Processing, and Handling of Combustible Particulate Solids*, provides design, inspection, and testing requirements for solids-handling systems. The design information includes criteria for (1) explosion and fire protection equipment, (2) control of explosion hazards, and (3) appropriate process equipment, such as material transfer systems (e.g., mechanical conveyors, pneumatic conveying systems), duct systems, pressure protection systems, air-moving devices, air-material separators, gates and dampers, size reduction devices, particle size separation equipment, mixers and blenders, and dryers.

NFPA 654 also includes inspection, testing, and maintenance guidelines for solids-handling systems. Specifically, the code instructs facilities to establish a program to inspect, test, and maintain the following:

- Fire and explosion protection and prevention equipment in accordance with the applicable NFPA standards
- Dust control equipment
- Housekeeping procedures

9. EQUIPMENT-SPECIFIC INTEGRITY MANAGEMENT                          157

- Potential ignition sources
- Electrical, process, and mechanical equipment, including process interlocks

In addition, NFPA 496 requires (1) lubricating material-feeding device bearings (e.g., conveyor drives), air-moving device bearings (e.g., fans, blowers), air-separation devices (as applicable), and gates/dampers, (2) periodically checking the bearings on material-feeding devices and air-moving devices for excessive wear, (3) periodically cleaning the material-feeding devices (if the conveyed material has a tendency to adhere to the device), (4) periodically checking air-moving devices for heat and vibration, (5) performing PM on fans and blowers, and (6) periodically inspecting the filter media on air-separation devices.

Additional guidance on preventing fires and explosions can be found in NFPA 69, *Standard on Explosion Prevention Systems*. The CCPS books, *Guidelines for Safe Handling of Powders and Bulk Solids*, and *Dust Explosions, Prevention and Protection; A Practical Guide*, also provide information on safe design and operation of solids-handling equipment.

### 9.8.4  Safety-critical Utilities

In some processes, loss of a utility can result in a process safety incident. For facilities with such processes, design, inspection, and testing activities for safety-critical utilities should be included in the MI program. The facility process hazard analysis (PHA) team should identify safety-critical utilities and specific safety-critical components of the utility system (assuming that the PHAs include evaluation of the loss of plant utility systems). In addition, PHA teams may consider the need to provide redundant or backup utility systems. Safety-critical utilities might include cooling water systems, electrical power, plant air, instrumentation, inert gas systems, plant water, and cooling systems for temperature sensitive materials.

The QA program for a safety-critical utility system should provide assurance of utility system and component reliability. Ensuring reliability often involves providing redundant equipment items (e.g., redundant cooling water supply pumps) and/or installing backup systems (e.g., emergency electrical generator). The ITPM activities for safety-critical utility systems typically include (1) function testing of the system redundancy and/or backup supply systems, (2) calibration and testing of instrumented systems used, and (3) performing tasks necessary to maintain the reliability of critical utility components (e.g., vibration analysis of cooling water pumps, infrared analysis of critical electrical switchgear).

### 9.8.5  Other Safety Equipment

A facility's MI program should also include safety showers, eyewash stations, employee alarm systems (e.g., evacuation alarms), and emergency response equipment (e.g., self-contained breathing apparatus [SCBA], firefighting equipment, spill containment supplies). The design, inspection, and testing requirements for much of this equipment are governed by ANSI standards (typically for the design), OSHA regulations, and NFPA codes. Table 9-12 provides a list of RAGAGEPs applicable to certain safety equipment. Typical ITPM activities for safety equipment include inventory maintenance, visual inspection, and function testing of the equipment.

TABLE 9-12
Summary of RAGAGEPs for Selected Safety Equipment

| Safety Equipment | Applicable Standard, Regulation, or Code | | |
|---|---|---|---|
| | ANSI | OSHA | NFPA |
| Eyewashes | ANSI Z358.1 | 29 CFR 1910.151 (c) | – |
| Safety showers | ANSI Z358.1 | 29 CFR 1910.151 (c) | – |
| Employee alarm systems | | 29 CFR 1910.165 | NFPA 72 |
| • Firefighting equipment<br>• Protective equipment<br>• SCBAs<br>• Fire apparatus | ANSI Z88.2 | 29 CFR 1910.156 | NFPA 600, 1851, 1911, 1915, 1971, 1981, and 1991 |
| Respiratory protection equipment | ANSI Z88.2 | 29 CFR 1910.134 | NFPA 1981 and 1991 |

## 9.9 EQUIPMENT-SPECIFIC MI ACTIVITY MATRICES

This section contains MI activity matrices for the following equipment:
- Pressure vessels (including columns, filters, and heat exchangers) (Table 9-13)
- Piping (Table 9-14)
- Pressure relief valves (Table 9-15)
- SISs and ESDs (Table 9-16)
- Pumps (Table 9-17)
- Fired heaters, furnaces, and boilers (Table 9-18)
- Switch gear (Table 9-19)

In addition, the CD accompanying this book contains MI activity matrices for other types of equipment.

## 9. EQUIPMENT-SPECIFIC INTEGRITY MANAGEMENT

### TABLE 9-13
### Mechanical Integrity Activities for Pressure Vessels

| New Equipment Design, Fabrication, and Installation | | Inspection and Testing | | Preventive Maintenance | | Repair | |
|---|---|---|---|---|---|---|---|
| Activity | Frequency | Activity | Frequency | Activity | Frequency | Activity | Frequency |

**Example Activities and Typical Frequencies**

| | | | | | | | |
|---|---|---|---|---|---|---|---|
| • Equipment specification, Vessel data sheet<br>• Process design requirements<br>• Materials selection<br>• Vendor/Shop qualification<br>• Equipment design by manufacturer<br>• Design approval by owner<br>• Welding/Quality control (QC) plan approval<br>• Equipment fabrication<br>• Inspection<br>• Documentation preparation<br>• Installation/Commissioning<br>• Acceptance and turnover | As required for fabrication and installation | External visual inspection | 5-year maximum | • Activities identified from RCM or similar work planning initiatives, such as:<br>• Routine visual surveillance<br>• Process conditions monitoring/tracking<br>• Process performance monitoring | As required to meet preventive maintenance schedule or process monitoring needs | • Equipment replacement-in-kind<br>• Unique vessel repair activities such as weld overlay, alterations, hot taps, or welding attachments to the pressure boundary<br>• Painting<br>• Insulation/Fireproofing repair<br>• Chemical cleaning<br>• Structural support and anchoring systems repair or renewal | As required by the condition of the equipment based on recommendations from ITPM activities or observations from normal operations |
| | | Thickness measurement | ½ corrosion life or 10-year maximum | | | | |
| | | Internal inspection or alternatively on-stream inspection (as applicable) | ½ corrosion life or 10-year maximum, thickness measurement suffices if corrosion rate is less than 5 mils per year | | | | |
| | | Additional inspections for specific degradation modes (e.g., corrosion under insulation) | As required by condition of equipment and rate of degradation | | | | |

## TABLE 9-13 (Continued)

| New Equipment Design, Fabrication, and Installation | Inspection and Testing | Preventive Maintenance | Repair |
|---|---|---|---|
| **Technical Basis for Activity and Frequency** | | | |
| QA practices for pressure vessels | Scheduled with intervals set by the results of previous activity or at fixed intervals based on inspection code (API 510 or National Board Inspection Code [NBIC]) or jurisdictional requirements | Company or jurisdictional requirements | Performed when indicated by failure during normal operations or by the results of ITPM activities |
| **Sources of Acceptance Criteria** | | | |
| ASME PV codes for design and fabrication, in conjunction with more stringent requirements in company engineering standards and in facility-specific or jurisdictional requirements for the pressure boundary | Acceptance criteria from inspection codes API 510, NBIC, and/or jurisdictional requirements. Acceptance criteria for damage from specific degradation modes per API RP 579 | Company requirements and good engineering practices, coupled with upper and lower safe limits for process conditions as defined in the process safety information (such as pressure, temperature, fluid composition, and velocity limits) | Design and fabrication codes: ASME PV codes, in conjunction with more stringent requirements in company engineering standards, or facility or jurisdictional requirements. In general, repairs and alterations are performed in accordance with ASME "R" stamp requirements |
| **Typical Failures of Interest** | | | |
| Incorrect material or weld metal, incorrect heat treatment, incorrect dimensions, misalignment or out-of-square flanges, leak during testing, weld defects, high hardness readings, use of unqualified welder or welding procedures | • Distortion of pressure boundary, leakage from cracks (e.g., fatigue, environmentally induced, stress corrosion cracking, caustic cracking), or holes in pressure boundary. Corrosion of pressure boundary, including corrosion under insulation<br>• Lack of grounding, and excessive corrosion of structural support and anchoring systems | • Distortion of pressure boundary, leakage from cracks (fatigue or environmentally induced), or holes in pressure boundary.<br>• Corrosion of pressure boundary, including corrosion under insulation. Lack of grounding, and excessive corrosion of structural support and anchoring systems | Incorrect material or heat treatment, incorrect dimensions, misalignment or out-of-square flanges, leak during testing, weld defects, high hardness readings, use of unqualified welder or welding procedures |
| **Personnel Qualifications** | | | |
| Company requirements and documented skills, NDE qualifications, inspection certifications or technical training for inspection and acceptance activities | Documented qualifications, industry inspection certifications (API 510 or NBIC), or specific technical training to analyze results | Tasks usually require craft-specific skills or operator-specific skills that are addressed within their respective training programs | Welders qualified per Section IX of the ASME Code. NDE technicians qualified in appropriate techniques. Industry inspection certifications (API 510 or NBIC) or specific technical training for pressure vessel engineering |

## 9. EQUIPMENT-SPECIFIC INTEGRITY MANAGEMENT

**TABLE 9-13** *(Continued)*

| New Equipment Design, Fabrication, and Installation | Inspection and Testing | Preventive Maintenance | Repair |
|---|---|---|---|
| **Procedure Requirements** | | | |
| Written procedures describing: <br>• Engineering standards for specification of equipment <br>• Project management (including hazard and design review schedules) <br>• Vendor qualification <br>• Documentation requirements <br>• Project acceptance and turnover requirements | Written procedures describing the inspection or test activity, including: <br>• The manner, the extent, the location, and date the inspection or test is performed and by whom <br>• The documentation and analysis of results <br>• The resolution of functions or condition not meeting acceptance criteria | These activities generally do not require task-specific procedures | • Craft skill procedures for typical tasks encountered in repairs (e.g., welding, gasket installation, bolt tightening, pressure testing) <br>• Job-specific procedures developed for repairs or alterations to the pressure boundary <br>• Job-specific procedures for unique or complex repairs or jobs with specialized technical content (e.g., retraying, modifications to internals, catalyst handling) <br>• Job-specific procedures with process engineering input for chemical cleaning |
| **Documentation Requirements** | | | |
| Company documentation requirements typically include U1 form, welding qualifications, design calculations, material certifications, QC results, heat treating records, as-built fabrication drawings, and nameplate rubbing | Results and analysis of each inspection are documented for the life of the equipment <br>• Inspection dates are tracked and technical deferral is required for late tests with alternate means of protection to be considered; deficient conditions are identified and resolved by the date recommended | Results are usually recorded by exception in equipment history files | Repair history is typically maintained with equipment inspection history |

## TABLE 9-14
## Mechanical Integrity Activities for Piping Systems

| New Equipment Design, Fabrication, and Installation | | Inspection and Testing | | Preventive Maintenance | | Repair | |
|---|---|---|---|---|---|---|---|
| Activity | Frequency | Activity | Frequency | Activity | Frequency | Activity | Frequency |

### Example Activities and Typical Frequencies

| Activity | Frequency | Activity | Frequency | Activity | Frequency | Activity | Frequency |
|---|---|---|---|---|---|---|---|
| • Design/fluid service requirements<br>• Pressure rating<br>• Materials selection<br>• Fabrication contractor qualification<br>• Design approval by owner<br>• Welding/QC plan approval<br>• Fabrication/storage/shipping<br>• Installation<br>• Acceptance inspection and testing<br>• Documentation preparation<br>• Acceptance and turnover<br>• Commissioning | As required for fabrication and installation | External visual inspection | Default interval values in API 570 | • Activities identified from failure modes and effects analysis (FMEA) or other analysis techniques for RCM, risk-based inspection (RBI), or similar work planning initiatives<br>• Process conditions monitoring/tracking | As required to meet preventive maintenance schedule | • Piping/component replacement-in-kind<br>• Commissioning activities<br>• Temporary clamps<br>• Hot taps/stopples, etc.<br>• Painting<br>• Insulation repair<br>• Cleaning<br>• Support, hanger and anchoring systems repair or renewal | As required by the condition of the equipment based on recommendations from the inspection and testing or preventive maintenance activities |
| | | Thickness measurement inspection | Lesser of default interval values in API 570 or half-life based on measured wall thickness and calculated corrosion rates | | | | |
| | | RBI assessment | Adjustment of intervals and extent with RBI assessment, plan to be reviewed at default inspection intervals | | | | |
| | | • Special emphasis inspection<br>• Injection point and soil-to-air interface | • Injection point inspection: Lesser of 3 years max. or half-life based on measured wall thickness and calculated corrosion rates<br>• Soil-to-air interface inspection: default interval values in API 570 | | | | |

## 9. EQUIPMENT-SPECIFIC INTEGRITY MANAGEMENT    163

### TABLE 9-14 (Continued)

| New Equipment Design, Fabrication, and Installation | Inspection and Testing | Preventive Maintenance | Repair |
|---|---|---|---|
| **Technical Basis for Activity and Frequency** | | | |
| Quality assurance practices for piping fabrication and installation | Scheduled with intervals set by the results of previous inspection or default maximum intervals listed in the inspection code (API 570) | Company or jurisdictional requirements | Performed when indicated by failure, by the results of preventive maintenance activities, by the results of inspection and testing activities |
| **Sources of Acceptance Criteria** | | | |
| ANSI/ASME B31 codes for piping design and fabrication and in conjunction with more stringent requirements in Company Engineering Standards or facility-specific standards | Acceptance criteria from inspection code API 570 or jurisdictional requirements. Acceptance criteria for damage from specific degradation modes per API RP 579. | Upper and lower safe limits for process conditions, such as pressure, temperature, fluid composition, and velocity, as defined in the process safety information | ANSI/ASME B31 Design and Fabrication Code; in conjunction with more stringent requirements in Company Engineering Standards, facility, or jurisdictional requirements. |
| **Typical Failures of Interest** | | | |
| Dimensional errors, incorrect material or weld metal, incorrect dimensions, misaligned or out-of-square flanges, incorrect pressure rating for a component, leak during testing, weld defects outside acceptance criteria, high hardness readings, use of unqualified welder or welding procedures | Leakage from cracks (e.g., fatigue, environmentally induced, stress corrosion cracking, caustic cracking), internal or external corrosion, corrosion under insulation, excessive vibration, unsupported or bound piping, permanent distortion, piping component not meeting pressure rating. | Process conditions exceed safe upper or lower limit | Dimensional errors, incorrect material or weld metal, incorrect dimensions, misaligned or out-of-square flanges, incorrect pressure rating for a component, leak during testing, weld defects outside acceptance criteria, high hardness readings, use of unqualified welder or welding procedure, leakage during hot tap/stopple operations, inability to remove hot tap/stopple machines |
| **Personnel Qualifications** | | | |
| Company requirements, and documented craft skills for installation, NDE qualifications, and ASME Section IX welding requirements for welders | Documented NDE qualifications, industry inspection certifications (API 570), or specific technical training for piping engineering for analysis of results. | | Welders qualified per ASME Section IX Code, NDE technicians qualified to appropriate techniques, industry inspection certifications (API 570), or specific technical training for storage tank engineering. |

## TABLE 9-14 (Continued)

| New Equipment Design, Fabrication, and Installation | Inspection and Testing | Preventive Maintenance | Repair |
|---|---|---|---|
| **Procedure Requirements** | | | |
| • Written procedures describing: <br> • Engineering standards for specification of equipment <br> • Project management (including hazard and design review schedules) <br> • Vendor qualification <br> • Documentation requirements <br> • Project acceptance and turnover requirements | • Written procedures describing the inspection or test activity that includes: <br> • The extent and location of the activity, how and when the inspection or test is performed, and by whom <br> • How the results are documented and when the results are analyzed <br> • How a function or condition not meeting the acceptance criteria is resolved | | • Craft skill procedures for typical tasks encountered in repairs, such as welding, gasket installation, bolt tightening, etc. <br> • Job-specific procedures developed for repairs or alterations to the pressure boundary <br> • Job-specific procedures for unique or complex repairs, or jobs with specialized technical content, such as line lifting, hot taps, stopples, and clamp installations <br> • Job-specific procedures with process engineering input for chemical cleaning |
| **Documentation Requirements** | | | |
| Company documentation requirements typically include welding qualifications, weld map, design calculations, material certifications, QC results, as-built drawings, and pressure test reports | • Results and analysis of each inspection are documented for the life of the equipment <br> • Inspection dates are tracked and technical deferral required for late tests with alternate means of protection to be considered; deficient conditions are identified and resolved by the date recommended | Results are usually recorded by exception in the equipment history file | Repair history is maintained with equipment inspection history |

# 9. EQUIPMENT-SPECIFIC INTEGRITY MANAGEMENT

## TABLE 9-15
## Mechanical Integrity Activities for Pressure Relief Valves

| New Equipment Design, Fabrication, and Installation | | Inspection and Testing | | Preventive Maintenance | | Repair | |
|---|---|---|---|---|---|---|---|
| Activity | Frequency | Activity | Frequency | Activity | Frequency | Activity | Frequency |
| **Example Activities and Typical Frequencies** | | | | | | | |
| • Design requirements and process specifications<br>• Component materials<br>• Sizing design basis and sizing calculations<br>• Vendor/shop qualification<br>• Equipment design by manufacturer<br>• Equipment fabrication<br>• Inspection and testing<br>• Documentation preparation<br>• Installation and commissioning<br>• Acceptance and turnover | As required for fabrication and installation | External visual inspection | Annual | • Activities identified from RCM or similar work planning initiatives, such as:<br>• Routine visual surveillance<br>• Process conditions monitoring/tracking | As required to meet preventive maintenance schedule or process monitoring needs | • Equipment replacement-in-kind<br>• Mounting locations - repair or renewal<br>• Piping conditions - internal restrictions<br>• Visual inspection after the PSV operates | • 5 years, or to date stamped on the nameplate<br>• As required by the condition of the equipment based on recommendations from ITPM activities or observations from normal operations<br>• API RP 576 |
| | | Process conditions, including positions of upstream and downstream valves | Weekly | | | | |
| | | Pop testing of pressure relief valve | As required by the service conditions | | | | |
| | | Inspection of inlet and outlet piping for fouling and plugging | Whenever the device is replaced or removed for testing | | | | |
| | | Additional inspections for specific degradation modes | As required by service conditions, condition of equipment, and rate of degradation | | | | |

## TABLE 9-15 (Continued)

| New Equipment Design, Fabrication, and Installation | Inspection and Testing | Preventive Maintenance | Repair |
|---|---|---|---|
| **Technical Basis for Activity and Frequency** | | | |
| QA practices for pressure relief valve fabrication, testing, and installation | Scheduled with intervals determined by the results of previous activities or at fixed intervals based on inspection codes (ASME, NBIC, or API RP 576) or jurisdictional requirements | Company or jurisdictional requirements | Performed when required by the qualification period for the device or as indicated by failure during normal operations or the results or ITPM activities |
| **Sources of Acceptance Criteria** | | | |
| • Codes and standards for design and fabrication of pressure relief devices (ASME BPVC- Section VIII, NB-23, NFPA 30, API RP 520, or other standards applicable for the specific application [e.g., ammonia, LPG]), in conjunction with requirements for pressure vessels<br>• Company engineering standards, facility-specific and/or jurisdictional requirements | • Acceptance criteria from inspection codes (ASME, NBIC, or API) or jurisdictional requirements<br>• Company standards for evaluating the condition of the device and process conditions | Upper safe limits for process conditions (pressure), company requirements, and good engineering practice for process conditions defined in the process safety information | • Design codes (ASME, NBIC, or API)<br>• Company engineering standards, facility or jurisdictional requirements<br>• Some jurisdictions require an ASME "VR" stamp for repairs |
| **Typical Failures of Interest** | | | |
| Incorrect materials or internal components, incorrect pressure rating for a component, weld defects outside of acceptance criteria, dimensional errors, misalignment or out-of-square flanges, leak during testing | • Leakage from the valve resulting from internal component fatigue, corrosion, leaking gaskets<br>• Failing to open at set pressure<br>• Plugging in the discharge piping by animals or water/ice resulting from loss of covers/flappers<br>• Process valves closed that prevent the device from functioning<br>• Fouling or pluggage of the vent header<br>• Failure of the inert gas purge system | • Leakage from the valve caused by failure to reset after functioning<br>• Process conditions in excess of the design criteria for the device<br>• Fouling or pluggage of the vent header<br>• Failure of the inert gas purge system | • Incorrect materials or internal components, incorrect pressure rating for a component, weld defects outside of acceptance criteria, dimensional errors, misalignment or out-of-square flanges, leak during testing<br>• Leakage from the valve caused by internal component fatigue or leaking gaskets |

## 9. EQUIPMENT-SPECIFIC INTEGRITY MANAGEMENT 167

### TABLE 9-15 *(Continued)*

| New Equipment Design, Fabrication, and Installation | Inspection and Testing | Preventive Maintenance | Repair |
|---|---|---|---|
| **Personnel Qualifications** | | | |
| • Manufacturer requirements, documented skills, inspection certifications, or technical training for inspection and acceptance activities during manufacture and installation<br>• Training on the sizing, selection, and specification of relief devices in accordance with applicable codes and standards | Specific technical training on pressure relief valve inspection, testing, handling, and installation procedures | Specific technical training on pressure relief valve inspection, testing, handling, and installation procedures | Specific technical training on pressure relief valve inspection, testing, handling, and installation procedures |
| **Procedure Requirements** | | | |
| • Written procedures describing:<br>• Engineering standards for specification of equipment<br>• Project management (including hazard and design review schedules)<br>• Vendor qualification<br>• Documentation requirements<br>• Project acceptance and turnover requirements<br>• Proper installation requirements | • Written procedures describing the inspection or test activity including:<br>• The manner, the extent, the location, and the date the inspection or test is performed and by whom<br>• The documentation and analysis of results<br>• The resolution of functions or conditions not meeting acceptance criteria | These activities generally do not require task-specific procedures | • Craft skill procedures for typical tasks encountered in repairs and replacements (e.g., gasket installation, bolt tightening, pressure testing)<br>• Job-specific procedures developed for repairs or replacements |
| **Documentation Requirements** | | | |
| Company documentation requirements typically include manufacturer's data forms, design and sizing calculations, material certifications, initial pop test results, QC results, and device drawings | • Results and analysis of each inspection are documented for the life of the equipment<br>• Inspection dates are tracked and technical deferral is required for late tests with alternate means of protection to be considered; deficient conditions are identified and resolved by the date recommended | Results are usually recorded by exception in the equipment history file | Repair history is maintained with equipment inspection history |

**168** GUIDELINES FOR MECHANICAL INTEGRITY SYSTEMS

TABLE 9-16
Mechanical Integrity Activities for SISs and ESDs

| New Equipment Design, Fabrication, and Installation | | Inspection and Testing | | Preventive Maintenance | | Repair | |
|---|---|---|---|---|---|---|---|
| Activity | Frequency | Activity | Frequency | Activity | Frequency | Activity | Frequency |
| Example Activities and Typical Frequencies | | | | | | | |
| Identification of materials of construction | When received | Calibration of field devices (e.g., transmitters, switches)<br>• Loop check<br>• Functional test<br>• Checking/running of logic solver diagnostics | As specified to meet the safety performance requirements | • Replacement of logic solver battery<br>• Replacement/cleaning of logic solver, operator interface, and engineering interface cabinet air filters | Manufacturers' recommendations | • Troubleshooting<br>• Replacement of field devices (e.g., transmitters, switches)<br>• Replacement of logic solver, operator interface, and engineering interface cabinet components (e.g., integrated circuit boards | As required |
| | | Calibration of field devices (e.g., transmitters, switches) | Before installation | | | Identification of materials of construction | When new devices are installed |
| | | Visual inspection of field devices and installation | Initial installation | | | | |
| | | SIS loop check | Initial installation | | | | |
| | | SIS functional test | Initial installation | | | | |
| | | Manufacturer testing of logic solver, operator interface, and engineering interface | During fabrication | | | | |
| | | Factory acceptance testing of logic solver, operator interface, and engineering interface | Before delivery and again after delivery | | | | |

## 9. EQUIPMENT-SPECIFIC INTEGRITY MANAGEMENT

### TABLE 9-16 (Continued)

| New Equipment Design, Fabrication, and Installation | Inspection and Testing | Preventive Maintenance | Repair |
|---|---|---|---|
| **Technical Basis for Activity and Frequency** | | | |
| • API RP 554<br>• API RP 551<br>• ISA S84.01 and/or IEC 61508<br>• Manufacturers' recommendations<br>• Industrial insurers' recommendations<br>• Common industry practices | • API RP 554<br>• API RP 551<br>• ISA S84.01 and/or IEC 61508<br>• Manufacturers' recommendations<br>• Industrial insurers' recommendations<br>• Common industry practices | Manufacturers' recommendations | Common repair activity |
| **Sources of Acceptance Criteria** | | | |
| • Device specifications<br>• SIS specifications<br>• Manufacturers' recommendations<br>• Industrial insurers' recommendations<br>• Company engineering and/or maintenance standards | • Device specifications<br>• SIS specifications<br>• Manufacturers' recommendations<br>• Industrial insurers' recommendations<br>• Company engineering and/or maintenance standards | Manufacturers' recommendations | • Device specifications<br>• SIS specifications |
| **Typical Failures of Interest** | | | |
| • Leakage, resulting from improper installation or materials of construction (e.g., incorrect gasket, incorrect metallurgy of wetted parts)<br>• Failure to operate on demand caused by improper installation (e.g., incorrectly wired), incorrect configuration of control system, and/or incorrect calibration of the device<br>• Potential ignition source or electrical shock to personnel as a result of improper installation<br>• Improper electrical classification rating for the electrical classification of the area where the device is installed | • Failure to operate on demand or spurious trip of system as a result of wiring failure (e.g., loose connection, short), failure of input device (e.g., sensor electronic failure), or failure of controller/local solver (e.g., I/O card failure)<br>• Failure to operate on demand or spurious trip as a result of an unauthorized change to controller/logic solver configuration and/or bypassing/forcing of the interlock/alarm | • Failure to operate on demand or spurious trip of system as a result of wiring failure (e.g., loose connection, short), failure of input device (e.g., sensor electronic failure), or failure of controller/local solver (e.g., I/O card failure)<br>• Failure to operate on demand or spurious trip as a result of an unauthorized change to controller/logic solver configuration and/or bypassing/forcing of the interlock/alarm | • Leakage at process connection, resulting from improper installation or materials of construction (e.g., incorrect gasket, incorrect metallurgy of wetted parts)<br>• Failure to operate on demand as a result of improper installation (e.g., incorrectly wired), incorrect configuration of control system, and/or incorrect calibration of the device<br>• Potential ignition source or electrical shock to personnel as a result of improper installation<br>• Improper electrical classification rating for the electrical classification of the area where the device is installed |

## TABLE 9-16 (Continued)

| New Equipment Design, Fabrication, and Installation | Inspection and Testing | Preventive Maintenance | Repair |
|---|---|---|---|
| **Typical Failures of Interest (Continued)** | | | |
| | • Failure to operate on demand caused by isolation/plugging of input device connection or other process conditions (e.g., buildup on a temperature probe) that render input device inoperable or incapable of accurately measuring process conditions (e.g., pressure, temperature)<br>• Potential ignition source if the device or connection shorts<br>• Leakage at process connection as a result of overpressurization of the joint | • Failure to operate on demand caused by isolation/plugging of input device connection or other process conditions (e.g., buildup on a temperature probe) that render input device inoperable or incapable of accurately measuring process conditions (e.g., pressure, temperature)<br>• Potential ignition source if the device or connection shorts<br>• Leakage at process connection as a result of overpressurization of the joint | |
| **Personnel Qualifications** | | | |
| • Craft skills and knowledge required by the individual procedures<br>• Training on the use and operation of special tools (e.g., signal simulators) required by procedures | • Craft skills and knowledge required by the individual procedures<br>• Training on the specific procedures for the inspection and testing activities<br>• Training on the use and operation of special tools (e.g., signal simulators) required by procedures | • Craft skills and knowledge required by the individual procedures | • Craft skills and knowledge required by the individual procedures<br>• Training on the specific procedures for the repair activities<br>• Training on the use and operation of special tools (e.g., signal simulators) required by procedures |
| **Procedure Requirements** | | | |
| • Procurement and receiving procedures to ensure proper materials of construction<br>• Device-specific testing, calibration, and installation procedures<br>• Manufacturers' manuals<br>• Special tool (e.g., signal simulator) use and operation procedures | • Written procedures describing the inspection or test activity, including:<br>• The manner, the extent, the location, and the date the inspection or test is performed and by whom<br>• The documentation and analysis of results<br>• The resolution of functions or conditions not meeting acceptance criteria | • Written procedures describing the inspection or test activity, including:<br>• The manner, the extent, the location, and the date the inspection or test is performed and by whom<br>• The documentation and analysis of results<br>• The resolution of functions or conditions not meeting acceptance criteria | • Generic written repair procedures for troubleshooting of SISs and replacement of integrated circuit boards that include references to the manufacturers' manuals<br>• Device-specific installation procedures<br>• Manufacturers' manuals<br>• Special tool (e.g., signal simulator) use and operation procedures<br>• Procurement and receiving procedures to ensure proper materials of construction |

## 9. EQUIPMENT-SPECIFIC INTEGRITY MANAGEMENT

### TABLE 9-16 (Continued)

| New Equipment Design, Fabrication, and Installation | Inspection and Testing | Preventive Maintenance | Repair |
|---|---|---|---|
| **Documentation Requirements** | | | |
| • Vendor material of construction reports<br>• Calibration record, including as-found and as-left conditions<br>• Loop check sheet, including as-found and as-left conditions<br>• Functional test record<br>• Manufacturers' test report<br>• Factory acceptance test report<br>• Facility acceptance test report<br>• Installation documentation to support pre-startup safety review (PSSR) requirements | • Calibration record<br>• Loop check sheet, including as-found and as-left conditions<br>• Functional test record<br>• Diagnostic check sheet | • Completed/closed work order<br>• Equipment PM record | • Completed/closed work order<br>• Work order or storeroom records of parts/materials used<br>• Return to service check sheet<br>• Vendor material of construction reports |

**TABLE 9-17**
**Mechanical Integrity Activities for Pumps**

| New Equipment Design, Fabrication, and Installation | | Inspection and Testing | | Preventive Maintenance | | Repair | |
|---|---|---|---|---|---|---|---|
| Activity | Frequency | Activity | Frequency | Activity | Frequency | Activity | Frequency |
| *Example Activities and Typical Frequencies* | | | | | | | |
| • Identification of materials of construction<br>• Performance test<br>• Pressure test | Initial fabrication | Visual inspection of sealing system | Each shift to weekly (depending on criticality) | Bearing housing and/or gearbox oil/lubricant level check | Each shift to weekly | • Mechanical seal replacement<br>• Pump disassembly and assembly | As required |
| | | Vibration analysis | • Continuous for pumps with large motors (e.g., 10,000 hp)<br>• Weekly to quarterly (depending on criticality and horsepower) | Changing of bearing housing and/or gearbox oil/lubricant | Manufacturers' recommendations | | |
| Alignment | Initial installation and any time components are loosened, removed, or replaced | Performance testing | Depends on service conditions and criticality | Analysis of bearing and/or gearbox oil/lubricant | Monthly to semi-annually (depending on criticality and history) | Identification of materials of construction | When parts are received and/or at time of installation |
| Rotational check | Initial installation and any time the driver is connected | Swapping of redundant pumps | Weekly to monthly | Lubrication of metal couplings (e.g., Falk Steelflex, Fast gear, or similar types) | Manufacturers' recommendations | | |
| Vibration analysis (baseline) | Initial startup | Operational/functional check of standby pump | Weekly to monthly | Internal inspection and rebuilding | As required, depending on history, service conditions, and criticality | | |

## 9. EQUIPMENT-SPECIFIC INTEGRITY MANAGEMENT 173

### TABLE 9-17 (Continued)

| New Equipment Design, Fabrication, and Installation | Inspection and Testing | Preventive Maintenance | Repair |
|---|---|---|---|
| **Technical Basis for Activity and Frequency** | | | |
| • Various codes, standards, or recommended practices from various organizations (e.g., API, Hydraulic Institute, ANSI, ISO) (application depends on industry and type of pump)<br>• Manufacturers' recommendations<br>• Common industry practices<br>• Industrial recommendations | • Manufacturers' recommendations<br>• Common industry practices<br>• Industrial recommendations | • Manufacturers' recommendations<br>• Common industry practices<br>• Industrial recommendations | Common repair activity |
| **Sources of Acceptance Criteria** | | | |
| • Applicable code, standard, or recommended practice<br>• Pump specifications<br>• Manufacturers' recommendations<br>• Company engineering and/or maintenance standards<br>• Industrial insurers' recommendations | • Pump specifications<br>• Manufacturers' recommendations<br>• Company engineering and/or maintenance standards | • Manufacturers' and/or lubricant/oil vendor's recommendations<br>• Company engineering and/or maintenance standards | • Pump specifications<br>• Manufacturers' recommendation<br>• Company engineering and/or maintenance standards |
| **Typical Failures of Interest** | | | |
| • Leakage from the seal/packing assembly as a result of improper installation or incorrect materials of construction<br>• Inadequate flow/pressure as a result of improper assembly or improper installation of internal components (e.g., inadequate impeller clearance) | • Leakage from the seal/packing assembly as a result of overpressurization of the assembly, inadequate lubrication, bearing failure, or wear<br>• Loss of, or inadequate, flow/pressure, resulting from failure of a drive component<br>• Loss of, or inadequate, flow/pressure, resulting from failure, corrosion, erosion, or wear of an internal component (e.g., impeller) | • Leakage from the seal/packing assembly as a result of overpressurization of the assembly, inadequate lubrication, bearing failure, or wear<br>• Loss of, or inadequate, flow/pressure, resulting from failure of a drive component<br>• Loss of, or inadequate, flow/pressure, resulting from failure, corrosion, erosion, or wear of an internal component (e.g., impeller) | • Leakage from the seal/packing assembly as a result of improper installation or incorrect materials of construction<br>• Inadequate flow/pressure resulting from improper assembly or improper installation of internal components<br>• Leakage from pump casing as a result of improper assembly, installation, or materials of construction |

## TABLE 9-17 (Continued)

| New Equipment Design, Fabrication, and Installation | Inspection and Testing | Preventive Maintenance | Repair |
|---|---|---|---|
| **Typical Failures of Interest (Continued)** | | | |
| • Leakage from pump casing as a result of improper assembly, installation, or materials of construction<br>• Damage to seal/packing and/or associated equipment caused by excessive vibration | • Damage to pump casing from loose or broken internal component (e.g., impeller)<br>• Damage to seal/packing and/or associated equipment caused by excessive vibration | • Damage to pump casing from loose or broken internal component (e.g., impeller)<br>• Damage to seal/packing and/or associated equipment caused by excessive vibration | • Damage to seal/packing and/or associated equipment caused by excessive vibration |
| **Personnel Qualifications** | | | |
| • Craft skills and knowledge required by the individual procedures<br>• Training on the specific procedures for the inspection and testing activities<br>• Training on the use and operation of special tools required by procedures (e.g., laser alignment equipment) | • Craft skills and knowledge required by the individual procedures<br>• Training on the specific procedures for the inspection and testing activities<br>• Training on the use and operation of special tools required by procedures (e.g., vibration analysis equipment) | • Craft skills and knowledge required by the individual procedures<br>• Training on the specific procedures for the preventive maintenance activities | • Craft skills and knowledge required by the individual procedures<br>• Training on the specific procedures for the repair activities<br>• Training on the specific procedures for the inspection and testing activities<br>• Training on the use and operation of special tools required by procedures (e.g., laser alignment equipment) |
| **Procedure Requirements** | | | |
| • Procurement and receiving procedures to ensure proper materials of construction<br>• Manufacturer testing procedures, including those for alignment tools<br>• Pre-commissioning and/or commissioning testing procedures<br>• Manufacturers' manuals<br>• Pump alignment procedure<br>• Pump installation procedure<br>• Vibration analysis procedure<br>• Vibration analyzer operation procedure and/or manufacturers' manual | • Written procedures describing the inspection or test activity including:<br>• The manner, the extent, the location, and the date the inspection or test is performed and by whom<br>• The documentation and analysis of results<br>• The resolution of functions or conditions not meeting acceptance criteria | • Written procedures describing the inspection or test activity including:<br>• The manner, the extent, the location, and the date the inspection or test is performed and by whom<br>• The documentation and analysis of results<br>• The resolution of functions or conditions not meeting acceptance criteria | • Generic written repair procedures for centrifugal pumps that include references to the manufacturers' manuals<br>• Pump alignment procedure<br>• Alignment tool (e.g., laser alignment) procedure and/or manufacturers' manual<br>• Pump installation procedure<br>• Procurement and receiving procedures to ensure proper materials of construction<br>• Testing procedures covering operation of testing equipment and/or performance of nondestructive testing |

## 9. EQUIPMENT-SPECIFIC INTEGRITY MANAGEMENT    175

### TABLE 9-17 *(Continued)*

| New Equipment Design, Fabrication, and Installation | Inspection and Testing | Preventive Maintenance | Repair |
|---|---|---|---|
| **Documentation Requirements** | | | |
| • Vendor material of construction reports<br>• Nondestructive testing records<br>• Performance and/or pressure test reports<br>• Alignment report<br>• Pump installation record<br>• Vibration analysis data and report | • Inspection check sheet<br>• Vibration analysis data and report<br>• Performance test records | • Lube route check sheet<br>• Completed/closed work order<br>• Equipment PM record<br>• Oil/lubricant analysis report | • Completed/closed work order<br>• Work order or storeroom records of parts/materials used<br>• Return to service check sheet<br>• Alignment report<br>• Pump installation record<br>• Vendor material of construction reports<br>• Nondestructive testing records |
| • Special notes:<br>• Records retained for the life of the equipment<br>• Installation documentation to support PSSR requirements | • Special notes:<br>• Records retained as required by the facility's record retention policy<br>• Documentation by exception may be acceptable (for certain tasks and select pumps) | • Special notes:<br>• Records retained as required by the facility's record retention policy<br>• Documentation by exception may be acceptable (for certain tasks and select pumps) | • Special notes:<br>• Repair data (e.g., condition found, parts used, repairs made, condition left) are usually recorded in the equipment history files<br>• Repair history retained for the life of the equipment |

# TABLE 9-18
## Mechanical Integrity Activities for Fired Heaters/Furnaces/Boilers

| New Equipment Design, Fabrication, and Installation | | Inspection and Testing | | Preventive Maintenance | | Repair | |
|---|---|---|---|---|---|---|---|
| Activity | Frequency | Activity | Frequency | Activity | Frequency | Activity | Frequency |
| **Example Activities and Typical Frequencies** | | | | | | | |
| • Heat or steam rate duty/fluid service requirements<br>• Pressure rating<br>• Materials selection<br>• Fabrication contractor qualification<br>• Design approval by owner<br>• Welding/QC plan approval<br>• Fabrication/storage/shipping<br>• Installation<br>• Acceptance inspection and testing<br>• Documentation preparation<br>• Acceptance and turnover<br>• Commissioning | As required for fabrication and installation | Inspection of heater or furnace firebox and tubes | Scheduled by user, usually in conjunction with planned unit shutdown or when an imminent failure condition is visible | • Activities identified from RCM or similar work planning initiatives, such as:<br>• Process conditions monitoring/tracking<br>• Metal temperature monitoring<br>• Water quality testing<br>• Efficiency performance monitoring | As required to meet preventive maintenance schedule and boiler operating permit requirements | • Tube replacement<br>• Burner replacement or repair<br>• Insulation/refractory repair<br>• Tube or piping support, hanger and anchoring systems repair or renewal<br>• Commissioning activities following repair | As required by the condition of the equipment, based on recommendations from inspection and testing or preventive maintenance activities |
| | | • Inspection of boilers, steam drums, tubes, and firebox<br>• Waste heat boiler inspection<br>• Hazardous waste incinerator inspection | Inspection schedule is not code derived, but is generally mandated by jurisdictional requirements | Decoking operations for heaters and furnaces | Performed when needed, based on performance and metal temperature data | | |

## 9. EQUIPMENT-SPECIFIC INTEGRITY MANAGEMENT 177

### TABLE 9-18 (Continued)

| New Equipment Design, Fabrication, and Installation | Inspection and Testing | Preventive Maintenance | Repair |
|---|---|---|---|
| **Technical Basis for Activity and Frequency** | | | |
| QA practices for equipment fabrication and installation | Jurisdictional requirements for boilers, and reliability issues for heaters and furnaces | Company or jurisdictional requirements | Performed when indicated by failure, or based on recommendations from preventive maintenance or inspection and testing activities |
| **Acceptance Criteria** | | | |
| • ASME and National Board codes for boiler design and fabrication, and installation<br>• Company engineering standards, or facility-specific standards for heaters and furnaces | • Acceptance criteria from NBIC or jurisdictional requirements<br>• Company standards for evaluating tube life and refractory damage | Upper and lower safe limits for process conditions defined in the process safety information, such as pressure, temperature, fluid composition, and velocity | • ASME and National Board codes for boiler repair<br>• Company Engineering Standards, or facility-specific standards for heaters and furnaces |
| **Typical Failures of Interest** | | | |
| Incorrect material or weld metal, incorrect pressure rating for a component, leak during testing, weld defects outside of acceptance criteria, high hardness readings, use of unqualified welder or welding procedures, dimensional errors, incorrect refractory material, incorrect refractory installation or cure, incorrect spring hanger setting, damage to finned tube | Tube ruptures, tube bulging and distortions, creep damage, steam leaks, process fluid leaks, refractory failure, structural support failure | Process conditions, such as outlet temperatures, stack temperatures, combustion efficiency, heat input, flow balances, and tubular meal temperatures, exceed safe operating limit | Incorrect material or weld metal, incorrect pressure rating for a component, leak during testing, weld defects outside of acceptance criteria, high hardness readings, use of unqualified welder or welding procedures, dimensional errors, incorrect refractory material, incorrect refractory installation or cure, incorrect spring hanger setting, damage to finned tube |
| **Personnel Qualifications** | | | |
| • Boilers: ASME and National Board fabrication requirements<br>• Heaters and furnaces: Company requirements, and documented craft skills for installation, NDE qualifications, and ASME Section IX welding requirements for welders | • National Board inspection certifications for boilers per jurisdictional requirements<br>• Optional industry certifications (API 570, 510) or company-specific training for heater and furnace inspection | Company requirements for tasks | • Boilers: ASME and National Board fabrication requirements<br>• Heaters and furnaces: Company requirements, and documented craft skills for repairs, NDE qualifications, and ASME Section IX welding requirements for welders, optional industry certifications (API 570, 510), or company-specific training for inspection of firebox piping |

## TABLE 9-18 (Continued)

| New Equipment Design, Fabrication, and Installation | Inspection and Testing | Preventive Maintenance | Repair |
|---|---|---|---|
| **Procedure Requirements** | | | |
| • Written procedures describing:<br>• Engineering standards for specification of equipment<br>• Project management (including hazard and design review schedules)<br>• Vendor qualification<br>• Documentation requirements<br>• Project acceptance and turnover requirements | • Written procedures describing the inspection or test activity, including:<br>• The manner, the extent, the location, and the date the inspection or test is performed and by whom<br>• The documentation and analysis of results<br>• The resolution of functions or conditions not meeting acceptance criteria | • Written procedures describing the inspection or test activity, including:<br>• The manner, the extent, the location, and the date the inspection or test is performed and by whom<br>• The documentation and analysis of results<br>• The resolution of functions or conditions not meeting acceptance criteria | • Craft skill procedures for typical tasks encountered in repairs (e.g., welding, gasket installation, bolt tightening)<br>• Job-specific procedures for unique or complex repairs, or for jobs with specialized technical content (e.g., coil or tube replacement, new spring hanger settings, refractory repairs or burner adjustments)<br>• Job-specific procedures with process engineering input for chemical cleaning |
| **Documentation Requirements** | | | |
| Company documentation requirements typically include: Manufacturers' data forms (boilers), welding qualifications, weld map, design calculations, material certifications, QC results, as-built drawings, and pressure test reports | Results and analysis of each inspection are documented for the life of the equipment | Results are usually recorded by exception in the equipment history file | Repair history is maintained with the equipment inspection history files |

## 9. EQUIPMENT-SPECIFIC INTEGRITY MANAGEMENT

### TABLE 9-19
### Mechanical Integrity Activities for Switch Gear

| New Equipment Design, Fabrication, and Installation | | Inspection and Testing | | Preventive Maintenance | | Repair | |
|---|---|---|---|---|---|---|---|
| Activity | Frequency | Activity | Frequency | Activity | Frequency | Activity | Frequency |
| *Example Activities and Typical Frequencies* | | | | | | | |
| • Visual inspection<br>• Verification/testing of overcurrent protection | Initial installation | Visual inspection | • Monthly to quarterly (outdoor installations)<br>• Quarterly to semiannually (indoor installation) | Overhaul of switch gear, including cleaning, inspecting, tightening, and adjusting of all components | 3 to 6 years, depending on conditions | Removal and installation of circuit breakers | As required |
| | | Infrared analysis | Annually to 3 years, depending on criticality, history, and starter size | | | | |
| | | Calibration and testing of protective relays; tripping of breakers; and testing insulation resistance of controls, meters, and protective devices | 3 to 6 years, depending on conditions | | | | |
| | | • Inspection and maintenance of circuit breaker<br>• Electrical testing of circuit breaker | 3 years maximum (air-break and oil-immersed circuit breakers) | | | | |
| | | System testing of installed circuit breaker | After completion of circuit breaker electrical testing | | | | |
| | | Manufacturers' recommended inspection, maintenance, and testing of vacuum and gas-filled circuit breakers | Manufacturers' recommendations | | | | |

## TABLE 9-19 (Continued)

| New Equipment Design, Fabrication, and Installation | Inspection and Testing | Preventive Maintenance | Repair |
|---|---|---|---|
| **Technical Basis for Activity and Frequency** | | | |
| • NFPA 70<br>• Manufacturers' recommendations<br>• Industrial insurers' recommendations<br>• Common industry practices | • NFPA 70B<br>• Manufacturers' recommendations<br>• Common industry practices<br>• Industrial insurers' recommendations | • NFPA 70B<br>• Manufacturers' recommendations<br>• Common industry practices | Common repair activity |
| **Sources of Acceptance Criteria** | | | |
| • NFPA 70<br>• Electric distribution system specifications<br>• Manufacturers' recommendations<br>• Industrial insurers' recommendations<br>• Company engineering and/or maintenance standards | • Electric distribution system specifications<br>• Manufacturers' recommendations<br>• Industrial insurers' recommendations<br>• Company engineering and/or maintenance standards | • Manufacturers/vendor's recommendations<br>• Industrial insurers' recommendations<br>• Company engineering and/or maintenance standards | • Electric distribution system specifications<br>• Manufacturers' recommendation<br>• Company engineering and/or maintenance standards |
| **Typical Failures of Interest** | | | |
| • Failure to provide electrical power to safety-critical devices (e.g., controls) as a result of incorrect installation (e.g., incorrectly connected or improper sizing of the unit)<br>• Potential ignition source or electrical shock to personnel as a result of improper installation or incorrect overcurrent protection device (e.g., incorrect fuse) | • Failure to provide electrical power to safety-critical devices (e.g., controls) as a result of failure of batteries, connection, transfer switch, and/or charging system<br>• Potential ignition source or electrical shock to personnel as a result of a short in device or connection, incorrect overcurrent protection device (e.g., incorrect fuse), or operating equipment with extreme load (e.g., the operating temperature is too high) | • Failure to provide electrical power to safety-critical devices (e.g., controls) as a result of failure of batteries, connection, transfer switch, and/or charging system<br>• Potential ignition source or electrical shock to personnel as a result of a short in device or connection, incorrect overcurrent protection device (e.g., incorrect fuse), or operating device with extreme load (e.g., the operating temperature is too high) | • Failure to provide electrical power to safety-critical devices (e.g., controls) as a result of incorrect installation (e.g., incorrectly connected or improper sizing of the unit)<br>• Potential ignition source or electrical shock to personnel as a result of improper installation or incorrect overcurrent protection device (e.g., incorrect fuse) |

## 9. EQUIPMENT-SPECIFIC INTEGRITY MANAGEMENT

**TABLE 9-19** *(Continued)*

| *New Equipment Design, Fabrication, and Installation* | *Inspection and Testing* | *Preventive Maintenance* | *Repair* |
|---|---|---|---|
| **Personnel Qualifications** | | | |
| • Craft skills and knowledge required by the individual procedures<br>• Training on the use and operation of special tools required by procedures (e.g., electrical test equipment) | • Craft skills and knowledge required by the individual procedures<br>• Training on the specific procedures for the inspection and testing activities<br>• Training on the use and operation of special tools required by procedures (e.g., electrical test equipment) | • Craft skills and knowledge required by the individual procedures<br>• Training on the specific procedures for the preventive maintenance activities | • Craft skills and knowledge required by the individual procedures<br>• Training on the specific procedures for the repair activities |
| **Procedure Requirements** | | | |
| • Manufacturers' manuals<br>• Electric distribution system installation procedures | • Written procedures describing the inspection or test activity including:<br>• The manner, the extent, the location, and the date the inspection or test is performed and by whom<br>• The documentation and analysis of results<br>• The resolution of functions or conditions not meeting acceptance criteria | • Written procedures describing the inspection or test activity including:<br>• The manner, the extent, the location, and the date the inspection or test is performed and by whom<br>• The documentation and analysis of results<br>• The resolution of functions or conditions not meeting acceptance criteria | • Generic written repair procedures for removal and installation of circuit breakers that include references to the manufacturers' manuals<br>• Procurement and receiving procedures to ensure the use of proper equipment |
| **Documentation Requirements** | | | |
| • Record of overcurrent protection devices (e.g., fuse ratings) and settings<br>• As-built drawings<br>• Installation documentation to support PSSR requirements | Inspection and testing check sheet | • Completed/closed work order<br>• Equipment PM record | • Completed/closed work order<br>• Work order or storeroom records of parts/materials used<br>• Return to service check sheet |

## 9.10 REFERENCES

9-1 Chlorine Institute, *Atmospheric Monitoring Equipment for Chlorine*, Pamphlet 73, Arlington, VA, 2003.

9-2 National Fire Protection Association, *Standard for the Installation of Oil-burning Equipment*, NFPA 31, Quincy, MA, 2001.

9-3 National Fire Protection Association, *National Fuel Gas Code*, NFPA 54, Quincy, MA, 2002.

9-4 National Fire Protection Association, *Boiler and Combustion Systems Hazard Code*, NFPA 85, Quincy, MA, 2004.

9-5 National Fire Protection Association, *Ovens and Furnaces*, NFPA 86, Quincy, MA, 2003.

9-6 National Fire Protection Association, *Standard for Industrial Furnaces Using a Special Processing Atmosphere*, NFPA 86C, Quincy, MA, 1999.

9-7 National Fire Protection Association, *Standard for Industrial Furnaces Using Vacuum as an Atmosphere*, NFPA 86D, Quincy, MA, 1999.

9-8 National Fire Protection Association, *Standard for Stoker Operation*, NFPA 8505, Quincy, MA, 1998

# 10
# MI PROGRAM IMPLEMENTATION

This chapter discusses issues related to implementing the mechanical integrity (MI) systems and activities previously discussed. Specifically, this chapter contains sections on the following topics:

- Budgeting and resources
- Use of software in MI programs
- Return on investment (ROI)

The budgeting and resources section outlines the resource needs for all phases of MI program implementation. The next section discusses the use of computerized maintenance management system (CMMS) software and other software in an MI program. Finally, the last section outlines some of the benefits a facility can expect from an effective MI program.

## 10.1 BUDGETING AND RESOURCES

Company and/or facility management often request an estimate of the resources needed to develop, implement, and sustain a successful MI program. This section provides information that can help personnel estimate these necessary resources.

Many companies use traditional project management tools and techniques to manage MI program development and implementation. For example, a project timeline depicted as a Gantt chart can be developed to document and communicate goals/targets for the activities and to monitor activity progress. Project cost monitoring and reporting systems can help track project costs. In addition, periodic project review meetings and status reports help keep efforts focused and provide opportunities to resolve issues as early as possible.

### 10.1.1 Program Development Resources

Company management should consider providing resources for the following efforts associated with defining and developing an MI program:

- Defining and documenting the overall management system for the MI program

- Identifying other MI program scope issues (e.g., covering program objectives, including nonregulated processes)
- Identifying equipment to include in the MI program
- Developing the inspection, testing, and preventive maintenance (ITPM) program and associated schedule
- Identifying and developing written procedures
- Defining the training program and developing and/or obtaining training program materials
- Developing a system for managing equipment deficiencies
- Defining quality assurance (QA) activities and writing QA procedures
- Defining any software needs and acquiring appropriate software

A primary resource for these activities is time: the time spent by personnel who are responsible for program activity, and the time given by other personnel who are needed to properly define and develop the program activities (e.g., craftspersons, inspectors, process engineers, operations personnel). These personnel resources can be obtained from in-house personnel and/or outside consultants experienced in MI program development.

The previous chapters of this book presented specific information associated with many of the topics listed above. Each of those chapters includes a section that defines typical roles and responsibilities. In addition, Table 10-1 provides an overview of resources and budgetary estimates of the labor time required for typical MI program development activities. Additional costs may be associated with specialized personnel training (e.g., inspector training), depending on the approach taken by the facility (e.g., whether or not to develop in-house expertise). Also, the estimates in Table 10-1 presume that the facility's process safety information (e.g., drawings, relief calculations) is accurate and up to date. The CD accompanying this book includes worksheets to use when identifying personnel needed to complete many of the development activities (Reference 10-1).

Outside consultants can be used to (1) provide technical expertise (e.g., knowledge of recognized and generally accepted good engineering practices [RAGAGEPs]), (2) facilitate the development of selected activities, (3) develop written MI procedures, and/or (4) augment the facility's staff when sufficient personnel are not available. While consultants can help with many development activities, successful MI programs require significant involvement of in-house personnel during the development phase. Often, MI program development efforts have been unsuccessful simply because they were contracted and did not include enough facility personnel involvement to achieve buy-in by plant personnel and to ensure that the program was practical for the specific culture, organization, and long-term resourcing for the facility. While consultants can be a valuable resource in developing an MI program (by providing direction, facilitating meetings, providing boilerplate program documents, etc.), MI programs developed primarily by consultants (i.e., with little involvement of facility personnel) may lack a level

## 10. MI PROGRAM IMPLEMENTATION

**TABLE 10-1**
**Summary of Resources Required for MI Program Development Activities**

| MI Program Development Activity | Typical Personnel Resources | | Labor Estimate (person-days) |
|---|---|---|---|
| | In-house | Outside | |
| Overall Management System | • MI coordinator<br>• Maintenance, Engineering, and Inspection manager(s)<br>• Maintenance management staff | Process safety consultant with MI program and reliability program experience | 10 to 25 |
| Other MI Program Scope Issues | • Plant management<br>• MI coordinator<br>• Maintenance and Engineering manager(s) | Process safety consultant with MI program and reliability program experience | 1 to 5 |
| Equipment List | • MI coordinator<br>• Process engineer(s)<br>• Maintenance management staff | Process safety consultant with MI program and reliability program experience | 5 to 30 |
| ITPM Program | • MI coordinator<br>• Maintenance, Engineering, and Inspection manager(s)<br>• Maintenance management staff<br>• Process engineer(s)<br>• Operations personnel | Process safety consultant with MI program and reliability program experience<br>Inspection company personnel | 10 to 75 |
| Procedures | • Maintenance management staff<br>• Craftspersons<br>• Inspection manager<br>• Inspectors<br>• Procedure writers | Procedure writers | 45 to 200 |
| Training | • Training department personnel<br>• Maintenance, Engineering, and Inspection manager(s)<br>• Maintenance management staff<br>• Craftspersons<br>• Inspectors | Training development consultants | up to 150 |

## TABLE 10-1 (Continued)

| MI Program Development Activity | Typical Personnel Resources | | Labor Estimate (person-days) |
|---|---|---|---|
| | In-house | Outside | |
| Equipment Deficiency | • MI coordinator<br>• Maintenance, Engineering, and Inspection manager(s)<br>• Inspectors<br>• Operations personnel | Process safety consultant with MI program and reliability program experience | 5 to 25 |
| QA Program | • Maintenance, Engineering, and Inspection manager(s)<br>• Inspectors<br>• Maintenance management staff<br>• Project engineering<br>• Purchasing<br>• Storeroom manager<br>• Procedure writers | • Process safety consultant with MI program and reliability program experience<br>• Procedure writers | 5 to 40 |
| Software | • Maintenance, Engineering, and Inspection manager(s)<br>• Inspectors<br>• Maintenance management staff<br>• Purchasing<br>• Storeroom manager | • Maintenance software vendors | 5 to 25 |

# 10. MI PROGRAM IMPLEMENTATION

of detail and specific process information needed for an effective MI program. A high level of involvement by facility personnel will help ensure that the consultants' "generic" programs contain and reflect the facility personnel's knowledge of the process and the facility's culture.

## 10.1.2 Initial Implementation Resources

During the initial implementation, the MI program moves from paper to the people and equipment in the field. This phase involves more time from the facility's personnel than the development phase did and is often the most costly phase. Typical activities for this phase are:

- Gathering and organizing equipment information
- Training and qualifying personnel who will perform MI activities, including obtaining required certifications (e.g., welding, inspector certifications)
- Implementing the ITPM tasks
- Managing ITPM results
- Implementing QA activities
- Managing equipment deficiencies
- Obtaining and implementing software to support the MI program

Table 10-2 provides a list of common tasks associated with each of these activities and an estimate of the labor time required to complete these activities. In addition, the following sections provide brief descriptions of these activities and discuss issues to consider when implementing these activities. Some of these initial implementation tasks will continue for the life of the MI program.

***Gathering and Organizing Equipment Information.*** Equipment information must be gathered and organized early in the implementation phase of the MI program. This information is needed during implementation to:

- Define acceptance criteria (usually, by referencing specific equipment file information [e.g., corrosion allowance, dimensional tolerances, acceptable wear]). See Sections 4.2.1 and 8.2 for additional information on acceptance criteria.
- Execute ITPM tasks. Specifically, personnel performing the inspection task should review selected equipment information with regard to (1) the ITPM history, (2) the equipment details (e.g., specific components requiring inspection), (3) the ITPM task details (e.g., thickness measurement locations [TMLs], inspection techniques), and (4) the acceptance criteria.

TABLE 10-2
Typical Initial Implementation Tasks by Activity

| Initial Implementation Activity | Typical Tasks | Estimated Labor Requirements |
|---|---|---|
| Gathering and Organizing Equipment Information | • Compile/organize equipment files<br>• Develop/obtain missing equipment information<br>• Establish/verify acceptance criteria for equipment (e.g., minimum wall thickness information, vibration tolerances) | 2 to 30 person-days<br>Highly variable, depending on the quantity of information to be developed/obtained<br>5 to 15 person-days |
| Training and Qualifying Personnel | • Develop a training plan<br>• Train maintenance craft personnel<br>• Qualify welders<br>• Certify inspectors<br>• Manage training efforts (e.g., maintain training database)[1] | 5 to 20 person-days<br>16 to 80 hours per craftsperson[2]<br>2 to 16 hours per welder[2]<br>24 to 80 hours per inspector[2]<br>1 to 5 person-days per month |
| Implementing ITPM Tasks | • Develop routes<br>• Develop inspection and testing schedule for individual equipment items<br>• Enter recurring work orders<br>• Perform route tasks[1] | 2 to 12 hours per route<br>1 to 4 hours per equipment item<br>0.5 to 4 hours per equipment item<br>1 to 4 hours per route |
| | • Perform ITPM tasks[1]<br>• Perform vessel/tank inspections<br>• Perform piping inspections<br>• Perform rotating equipment inspections<br>• Perform instrument/interlock testing | 2 to 8 hours per inspection<br>1 to 12 hours per circuit<br>1 to 4 hours per rotating equipment item<br>1 to 4 hours per instrument/interlock |

## 10. MI PROGRAM IMPLEMENTATION

**TABLE 10-2** *(Continued)*

| Initial Implementation Activity | Typical Tasks | Estimated Labor Requirements |
|---|---|---|
| Managing ITPM Results | • Implement inspection and testing results software<br>• Review results[1]<br>• Route results<br>• Review individual equipment inspection and testing results (e.g., inspection reports) and perform any needed calculations (e.g., remaining life estimates) | Highly variable, depending on software<br>0.5 to 4 hours per route<br>0.5 to 2 person-days per inspection/test report |
| Implementing QA Activities | • Train personnel (e.g., purchasing, project personnel, storeroom)<br>• Perform storeroom QA tasks[1]<br>• Perform project QA tasks[1]<br>• Review documents (e.g., specifications)<br>• Inspect major equipment items<br>• Verify installation (i.e., field verification) | 2 to 24 hours per individual<br>0.5 to 2 hours per task execution<br>0.5 to 4 hours per document<br>1 to 8 hours per inspection<br>Highly variable depending on project size |
| Managing Equipment Deficiencies | • Pilot test equipment deficiency resolution process<br>• Develop equipment deficiency tracking system<br>• Conduct equipment deficiency status review meetings[1] | 1 to 5 person-days<br>1 to 5 person-days<br>1 to 2 hours per meeting |

[1] Indicates that the task is an ongoing task as well as an initial implementation task
[2] Estimate is based on craftspersons and inspectors who are proficient in required basic job skills

The equipment information is usually located in the site's equipment files. The information in these files varies, depending on the type of equipment and the facility's recordkeeping culture and philosophy. Typically, equipment files contain such information as:

- Equipment design and construction data, such as design codes and standards used, design specifications, as-built drawings, materials of construction, dimensions (e.g., wall thickness, impeller diameter), and performance data (e.g., pressure relief valve [PRV] settings).
- Service history, such as length of time in service, materials handled, and changes in service
- ITPM history
- Maintenance history (i.e., failure history)
- Vendor-supplied information, such as installation instructions, dimensional specifications and allowable tolerances (e.g., the diameter of a journal or shaft), bolt materials and torque requirements, gasket and o-ring material requirements, lubricant specifications, maintenance and operating instructions, testing and maintenance recommendations, and performance testing data (e.g., pump performance testing)

This information may also be found in the facility's process safety information. Any necessary information not available at the site will need to be obtained from the equipment manufacturers. If the manufacturer is unable to provide the necessary information, facility personnel may have to develop the information or contract with appropriate engineering consultants to develop the information. Table 4-1 (in Chapter 4) provides a list of information needed for selected types of equipment.

***Training and Qualifying Personnel.*** As outlined in Chapter 5, personnel performing ITPM tasks and other MI tasks (e.g., equipment repairs) need to receive training on a variety of topics and may need to obtain certifications before performing certain tasks (e.g., pipe welding in accordance with American Society of Mechanical Engineers [ASME] B31.3). Facilities must also train personnel on applicable job procedures (e.g., ITPM procedures, repair procedures, QA procedures). For training to be effective, facilities are challenged to (1) identify training methods that are efficient and that keep experienced workers engaged and (2) allocate time for the trainers and trainees to be away from their regular job duties. To help overcome these issues, facilities can employ a variety of training methods (see Section 5.2 for more information on training methods) and should develop a training plan that accounts for personnel scheduling issues. This plan can then be used to estimate the resources required to complete the initial training. Training resources will typically include:

- Overtime to cover the absence of personnel from their regular duties to attend and perform the training
- Costs for outside trainers and/or training materials

- Administrative time to manage the training plan implementation and maintain the training records

Contractors at the facility will also require training. While this activity may not be as resource intensive as training facility personnel, resources should be allocated to:

- Obtain and review contractor safety and capabilities information
- Train contract employer representatives and/or contract employees
- Review contractor training records
- Ensure that only trained contract employees are used to perform work

***Implementing ITPM Tasks and Managing ITPM Task Results.*** The most resource-intensive activity during initial implementation of the MI program is likely the implementation of the ITPM tasks. Implementation of ITPM tasks typically begins with:

- Organizing the frequently performed tasks that can be performed on numerous pieces of equipment at a single time into maintenance rounds (e.g., lubrication rounds, vibration analysis rounds)
- Developing the inspection and testing schedules for tasks that are performed on individual equipment items
- Entering recurring work orders into the CMMS

The facility staff should consider the following factors when developing a schedule for maintenance rounds and individual ITPM tasks:

- Manpower loading. Performing all weekly activities on the same day of the week or all monthly activities on the first of the month is not practical or advisable. The facility staff should devise schedules that spread tasks out over the course of the week and/or month.
- Availability of qualified manpower to perform the activities.
- Operational issues, such as which pieces of equipment can or cannot be taken out of service at the same time and the ability to take equipment out of service (e.g., storage tanks for internal inspection, relief valve removal and testing, safety instrumented system [SIS] and interlock testing).

The maintenance planning and scheduling group at most facilities provides much of the personnel needed to complete the schedule; however, production management must be involved to ensure effective communication between production and maintenance personnel regarding the scope and any associated equipment downtime requirements of any maintenance task or activity to be accomplished. If outside contractors are used for certain ITPM tasks (e.g., American Petroleum Institute [API] 510 inspections of pressure vessels), consider involving them in the scheduling process or asking them to develop a proposed schedule and budget.

Once defined, the schedule can be used to identify the required resources and to develop a budget for implementing the tasks. In developing the budget, the following issues should be considered:

- Internal staffing requirements for routes and individual ITPM tasks
- Need for, and use of, contractors
- Required equipment outages
- Turnaround schedules

In addition to the resource and budget requirements directly associated with the ITPM schedule, the following issues should be considered when developing a budget for the implementation of ITPM tasks:

- Obtaining special equipment used to perform some tasks (e.g., vibration monitoring equipment, ultrasonic thickness [UT] measurement instruments)
- Obtaining and implementing software needed to manage the ITPM program and task results
- Costs associated with preparing equipment for inspection (e.g., coring pipe insulation)
- Equipment/system changes needed to maintain efficient performance of ITPM tasks (e.g., installation of block valves under pressure instruments)

Resource requirements for managing ITPM task results include time allocation for assigned personnel to review the results and determine what corrective actions are needed. This can include (1) a few hours per week for the maintenance craft supervisor to review the results from weekly routes and (2) a significant number of hours for engineers to review the numerous inspection reports. In addition, managing ITPM results may require purchasing special software to collect and analyze the volumes of data generated. (See Section 10.2 for a discussion of issues related to the need for, and use of, software in an MI program.)

***Implementing QA Activities.*** The resources required to implement the QA activities will be a function of the QA plan that is established. These resources may include:

- Acquiring RAGAGEP documents and training personnel to use them effectively
- Implementing vendor and contractor QA plans
- Obtaining tools for positive material identification (PMI)
- Providing space for receiving and storage, and furnishing the facilities/equipment needed to maintain required storage conditions (e.g., humidity and/or static control)
- Providing personnel training on receiving and storage procedures

## 10. MI PROGRAM IMPLEMENTATION

*Managing Equipment Deficiencies.* Initial implementation costs may be associated with managing equipment deficiencies. The resources for managing equipment deficiencies include the personnel and time needed to (1) develop and document the corrective actions required to bring the equipment back to its original specifications/condition and/or implement temporary corrective actions (if appropriate), (2) track equipment deficiencies to resolution, and (3) document final resolution. When initially implementing an equipment deficiency resolution program, resources can be used to:

- Pilot test the process
- Develop an equipment deficiency tracking system on the CMMS or other database
- Implement periodic equipment deficiency review meetings to ensure that deficiencies are being resolved in a timely manner

Because the management of change (MOC) program may be used to manage equipment deficiencies, involving personnel who are responsible for the MOC program would be helpful in this effort.

*Obtaining and Implementing Software.* Most facilities use software to implement and maintain the MI program. This includes software to help manage (1) the ITPM task schedule and work orders, (2) some ITPM task data (e.g., thickness measurements), (3) the MI program documents (e.g., procedures), (4) the MI training activities, and (5) corrective action tracking activities. Section 10.2 provides additional information on the use of software in the MI program.

During the MI program implementation phase, facility personnel need to (1) identify software needs, (2) assess any existing software's ability to satisfy these needs, (3) modify existing software and/or obtain additional/new software if necessary, (4) implement any new software or modifications to the existing software, and (5) provide training on any new or modified software systems. Maintenance software vendors/consultants can be a helpful resource during this phase.

### 10.1.3 Ongoing Efforts

Ongoing efforts for the MI program involve both the continued execution of the initial implementation tasks listed in Table 10-2 as well as the inclusion of new tasks to be performed. These new tasks focus on:

- Providing ongoing training
- Maintaining and improving existing MI procedures and developing new procedures as necessary
- Optimizing the ITPM tasks
- Maintaining QA activities
- Managing program changes

The initial development of the training program typically receives sufficient attention; however, facilities sometimes overlook the ongoing training effort that will be necessary. Allocating resources and budgets for ongoing training is important to the overall success of the MI program. Ongoing training efforts include refresher training and training on new topics. Many organizations provide annual refresher training that focuses on regulatory-required training, required certification/qualification training, and safe work practices. However, many organizations have expanded annual refresher training to include reviews of (1) job tasks and procedures that are of high risk or are performed infrequently and (2) procedures that have proven to be troublesome (i.e., tasks that are often not properly executed). The ongoing training budget should also recognize training needs for new employees and for reassigned/promoted employees. New training topics for the existing workforce may include new equipment or new maintenance/inspection techniques.

Trevor Kletz, author of *What Went Wrong: Case Histories of Process Plant Disasters*, wrote, "Procedures are subject to a form of corrosion more rapid than that which affects the steelwork; they vanish without a trace once management stops taking an interest in them..." (Reference 10-2). Without ongoing procedure development/modification efforts, the value of the initial procedure development effort will quickly be lost. Ongoing procedure development resources may be needed to update and correct procedures as procedure users discover errors or suggest improved execution steps. Also, facilities need to perform periodic reviews of both the procedures, to ensure that they stay current and accurate, and the personnel performance, to ensure that the procedures are followed. In addition, new procedures will be needed as new tasks are identified and implemented.

Some ITPM task optimization will result from reviewing task results (and performing remaining life or similar time-to-retirement calculations, if applicable). Optimization can result in changing the task activities and/or the task frequency. ITPM task optimization may also involve applying risk management tools to better understand potential failures (or root cause analysis [RCA] of actual failures) so that appropriate ITPM tasks can be planned. Chapter 11 discusses the most common risk management tools and Chapter 12 includes a section on RCA. Providing the resources and budget to optimize ITPM tasks is essential to realize some of the benefits of the MI program, such as:

- Improved safety and environmental performance
- Extended life of equipment (e.g., by identifying needed repairs before damage progresses to a level that makes repair impractical)
- Better capital planning (because the need to replace major equipment items is better predicted)
- Improved equipment reliability, including increased on-stream performance and the resulting economic benefits
- Improved ITPM task effectiveness and efficiency

## 10. MI PROGRAM IMPLEMENTATION

Resources from various departments/organizations within a company are needed to maintain QA activities. To ensure that the QA activities are sustained, these resources should be identified and responsibilities should be incorporated into the job duties for the involved job positions. Table 10-3 outlines some of the ongoing resources needed from various organizations.

TABLE 10-3
Examples of Ongoing QA Activities

| QA Area | Example QA Activities | Personnel Involved |
|---|---|---|
| Project Design | • Specification reviews<br>• Design reviews | • Project engineers<br>• Process engineers<br>• Equipment engineers<br>• Materials engineers<br>• Operations personnel |
| Vendor Selection and Management (both project and maintenance materials) | • Vendor qualification reviews<br>• Vendor audits<br>• Approved vendor list updating<br>• Vendor performance monitoring | • Purchasing personnel<br>• Project engineers<br>• Maintenance personnel<br>• Storeroom personnel |
| Contractor Selection and Management (both project and maintenance materials) | • Contractor safety performance and program reviews<br>• Contractor capability reviews<br>• Approved contractor list updating<br>• Contractor safety training sessions for contractor representatives and/or employees | • Purchasing personnel<br>• Project engineers<br>• Maintenance personnel<br>• Safety department personnel |
| Equipment Fabrication (off site) | • Vendor equipment drawing reviews<br>• Fabricator site visits<br>• Equipment inspections and tests | • Project engineers<br>• Outside nondestructive testing (NDT) contractors<br>• Inspectors |
| Equipment Fabrication (on site) and Installation | • Equipment inspections and tests<br>• Walkthroughs, including piping and instrumentation diagram (P&ID) verification<br>• Installation practices (e.g., equipment alignment)<br>• Onsite contractor safety and QA audits<br>• Equipment precommissioning activities | • Project engineers<br>• Outside NDT contractors<br>• Inspectors<br>• Operations personnel<br>• Maintenance craftspersons |
| Spare Part and Maintenance Material Procurement, Receiving, Storage, and Issuing | • Spare part information updates, including purchasing information<br>• Receiving practices (e.g., matching paperwork, tagging items)<br>• Receipt inspections and tests (e.g., PMI, visual inspection)<br>• Proper storage of items with specific storage conditions<br>• Storeroom inventory control practices and audits<br>• Issuing practices and inspections | • Purchasing personnel<br>• Inspectors<br>• Maintenance personnel<br>• Storeroom personnel |

Some of the resources listed in Table 10-3 may be provided from outside the facility, such as corporate purchasing and project engineering. Experience has shown that outside resources, especially for capital projects, typically require more oversight than facility-provided resources. This may result because:

- Outside personnel interacting with the facility personnel often change (e.g., different project managers).
- The outside personnel may not be as familiar with the facility's MI program processes and procedures.
- The goals, objectives, and expectations for personnel from outside the facility may not include issues related to process safety and the MI program; therefore, these individuals may not be aware that their job functions impact the process safety and MI programs. For example, corporate project engineers need to understand that the MI program can affect (and assist with) such issues as contractor selection, equipment fabrication, equipment installation, and project startup.

To help minimize these issues, facility management should consider (1) involving personnel familiar with the QA requirements of the MI program early in the project life cycle, (2) communicating the MI program's QA requirements to outside personnel, and (3) informing and/or training outside personnel on the QA program policies and procedures. Early involvement of MI program personnel can help ensure that the QA activities are fully and effectively integrated into the project. In addition, facilities can provide facility representatives to support outside project engineering and to help ensure that facility QA requirements are addressed.

Periodically, changes to the MI program will necessitate the allocation of resources to maintain the program. Some of the typical program changes are:

- Organizational changes. If a facility changes its organizational structure, MI program responsibilities may need to be changed.
- Process changes. Process changes can result in changes to the ITPM tasks and the associated task procedures and schedule.
- Equipment changes. As equipment is added or replaced, the ITPM tasks and equipment information may need to be updated.

## 10.2 USE OF SOFTWARE IN MI PROGRAMS

While successfully managing an MI program using paper-based systems is possible, most facilities find that managing many of the MI program activities is not practical without computers. Typically, the most important computer system used in MI is the facility's CMMS; however, other MI computer applications are also in use.

## 10. MI PROGRAM IMPLEMENTATION

### 10.2.1 Use of CMMS

Significant differences in CMMS software capabilities exist; however, almost all CMMS software packages contain the basic functions needed for an MI program. Two of the most important features are the CMMS work scheduling function and the work order function. The CMMS should be able to track when an ITPM task was last performed, calculate the date that the task should be repeated, and generate a timely work order.

In addition, the CMMS is typically used to manage the individual ITPM tasks and the equipment repair and replacement tasks. For individual ITPM tasks, the generation of a work order is typically the trigger that begins the scheduling process for the task. In addition, the work order can provide or reference the information needed to execute the task, such as equipment item, ITPM task description, ITPM task procedure, and locations of other pertinent information. Completed work orders can be used to document some ITPM task results: the date the task is completed, the name of the individual performing the task, a description of the task performed, an equipment identifier (e.g., tag number, asset number), and the results of the task. However, many CMMS programs have limited ability to accommodate the volume of data generated during some ITPM tasks (e.g., thickness measurement data, vibration analysis readings), and some programs may be unable to perform necessary calculations (e.g., corrosion rate, remaining life). Therefore, facilities frequently supplement the CMMS with additional software specifically to manage the information from these data-intensive ITPM tasks (see the following section for more information on these software packages).

For repair/replacement tasks, the CMMS can provide such information as the procedure for the task, a reference to other documents needed to perform the task (e.g., manufacturer's manual, permit requirements), and the means to manage equipment deficiencies. Work orders or work order logs can be used to communicate equipment deficiencies to affected personnel. (Note that the CMMS may need to be supplemented by other forms of communication.) The CMMS may be used to (1) identify and document corrective actions, including approval of temporary corrective actions, (2) track deficient equipment until permanent repairs are implemented, and (3) document the resolution and correction of the equipment deficiency.

A CMMS is also frequently used to assist facilities with the QA of spare parts and maintenance materials. Most systems have the following capabilities, which can help ensure that only correct parts and materials are used:

- Controlling and generating purchasing information to ensure that the correct spare parts and maintenance materials are ordered.
- Including notes (or other information) in the purchasing and/or spare parts information that will communicate specific QA requirements to appropriate personnel (e.g., vendors, receiving personnel).
- Integrating spare parts information with work orders to ensure that correct spare parts are ordered from the maintenance storeroom.

- Monitoring storeroom inventory to help ensure that appropriate parts are available.
- Tracking the use of spare parts and maintenance materials.

Additional CMMS features that are helpful for managing an MI program are failure coding, cost tracking, and report generation. CMMS programs frequently have the capability to enter failure codes for work orders. An effective failure coding system can identify equipment types (e.g., specific pump type) or specific equipment items with repeat failures that require further analysis (e.g., root cause or failure analysis; see Chapter 11 for more information on these analysis techniques). Also, CMMS programs can often record labor and material cost data associated with each work order. This cost information can be used in the performance measurement system and other aspects of MI program management. Finally, the CMMS should be able to provide MI program management reports, such as: MI-covered equipment lists, planned/scheduled ITPM task lists, completed ITPM task lists, overdue ITPM task reports, and equipment deficiency status reports.

### 10.2.2 Other Software Used in MI Programs

In addition to the CMMS, other software packages can be used to manage specific aspects of the MI program. Some of these packages are (1) software for specific ITPM tasks, (2) training management software, (3) document management software, and (4) software for risk management activities associated with the MI program. The following sections briefly describe each of these types of software.

***ITPM Task Data Collection and Analysis Software.*** Specialized software packages are used for some ITPM tasks to provide an interface to data-collection devices and to manage the quantity of data generated. Nondestructive examination (NDE) techniques, such as eddy current and UT measurement, often use specialized software packages to retrieve the data from field data-collection devices and to manage the large quantity of data generated. This software is typically used to (1) document the data so that reports can be generated, (2) highlight readings outside of acceptable limits, and (3) perform relevant calculations.

Specialized software for instrumentation and vibration analysis provides similar functions for instruments and rotating equipment. Instrument calibration software typically contains a catalogue of the instruments with associated information (e.g., model number, tag number, calibration range, calibration tolerance). This software (1) communicates with the devices used to calibrate the instruments and (2) helps manage the calibration data with features to document data, generate reports, identify out-of-tolerance readings, and perform error calculations. In addition, online instrument software is becoming increasingly common. Online software can identify malfunctioning instruments (e.g., transmitters, valves) before a failure affects system performance. Similarly, some vibration analysis software communicates with field vibration analyzers to collect

## 10. MI PROGRAM IMPLEMENTATION 199

and analyze vibration readings. Usually, the vibration software can document the data, generate reports, and identify readings that might indicate the onset of failures in the rotating equipment. Computer applications are continually evolving as developers add more features that can be used to support MI programs (e.g., compressor performance analysis).

***Training Program Software.*** Another type of software that can assist MI program management is training database software. Usually, this software provides means to document the training requirements for employees and can manage individual employee training records. These software packages often contain features to define the initial training requirements for new employees (or employees transferring to a new position) and refresher training requirements for the current workforce. The software should be able to generate a training plan, usually on an annual basis, and a training schedule for each employee in the organization. In addition, such software commonly includes features to document the training that each employee receives and maintains that information in individual training records. These records typically document the training topic, the date of training, the duration of the training, the means used to verify that the employee understood the training, and whether the employee successfully completed the training. In addition, these software packages can generate a variety of reports, such as reports to verify that employees have completed the training required to perform a task or to provide evidence of employee training during audits. Frequently, facilities have other training programs (e.g., operator training, safety training) that may provide database resources for the MI program.

***Document Management Software.*** Document management software can be very helpful for many maintenance organizations, especially for those with no experience managing the number of written procedures needed for an effective MI program. These software packages provide document organization and retrieval tools often used to manage MI procedures, equipment file information, and manufacturers' manuals. MI programs must ensure that up-to-date ITPM procedures, repair/replacement procedures, equipment file information, and manufacturers' manuals are available and accessible to the maintenance personnel and inspectors. This can be difficult and cumbersome using a paper-based system. In addition, paper-based systems risk losing or damaging unique paper copies of equipment file information and manufacturers' manuals. Document management software allows personnel to maintain procedures, equipment file information, inspection reports, and manufacturers' manuals in an electronic format that can be retrieved and printed when needed (e.g., when issuing a work order). In addition, some of these software packages can work in conjunction with CMMS programs so that required documents can be printed with the work order.

***Risk Management Software.*** Chapter 11 discusses several risk management activities that personnel can implement to improve and optimize aspects of the MI program. Most of the activities focus on improving ITPM tasks. The most common risk management methods used in conjunction with the MI program are failure modes and effects analysis (FMEA), risk-based inspection (RBI), and reliability-centered maintenance (RCM) analyses.

Several software packages are available to help employees perform and document the results of these activities. This type of software is used to assist the study leader in structuring an analysis and to provide specific analysis tools (e.g., lists of failure modes, criteria for assessing damage mechanisms). Frequently, such software can be used to document results and generate study reports. In addition, many of these software packages, especially RBI software, include features to update the analysis and the analysis results based on the completed ITPM task results.

## 10.3 RETURN ON INVESTMENT (ROI)

An MI program requires considerable resources; most managers are interested in the ROI of these resources. An MI program can expect benefits in the following areas:

- Equipment reliability
- Cost avoidance (including safety, environmental, and financial costs)
- Regulatory compliance and industry association commitments
- Reduced liability and reduced damage to corporate reputation

The ROI for MI programs depends on many factors, such as the maintenance systems in place prior to initial MI program development and implementation, the size of the organization, and the number of outside resources needed. Quantifying the overall ROI for an MI program can be problematic because of the difficulties in measuring some of the MI program benefits. For example, determining and/or estimating the benefits associated with preventing a fire in a facility, complying with government regulations, and/or maintaining market share (i.e., avoiding lost market share as a result of negative publicity associated with a catastrophic event) can be very difficult. The following paragraphs briefly describe the benefits and resulting ROI for each of the above-mentioned areas.

### 10.3.1 Improved Equipment Reliability

One of industry's primary objectives of the MI program is to replace the "breakdown" maintenance philosophy with a more proactive maintenance philosophy (Reference 10-3). The basic purpose of the ITPM program is to define and implement tasks/procedures that will detect the onset of equipment failures and/or prevent equipment failures. For many facilities, this shift in philosophy can result in significant improvements in equipment reliability. In addition, an effective MI program improves a facility's ability to predict when equipment repair or replacement is needed. This increased understanding of equipment condition allows the facility to better plan repairs and replacement, which reduces the impact of unexpected equipment failures.

## 10. MI PROGRAM IMPLEMENTATION

In addition, the QA program supports improved equipment reliability with activities that help ensure the suitability of the initial design, fabrication, and installation of process equipment. While these activities may result in some additional upfront costs, they are vital to ensuring equipment reliability over the equipment's life. For example, installing equipment with the wrong materials of construction or forgoing installation standards (e.g., rotating equipment alignment standards) will likely result in premature failures that can lead to both equipment repair and lost production costs.

Furthermore, the equipment deficiency resolution process helps ensure that (1) equipment breakdowns are properly managed and (2) temporary repairs are tracked until they are permanently corrected. (Historically, many temporary repairs have been forgotten until failures occurred that were worse than the initial failures.) In addition, facilities can employ equipment deficiency programs to identify chronic failures that can then be subjects for RCA and subsequent correction. These activities are aimed at improving equipment reliability, thus reducing unplanned downtime associated with equipment failures, as well as operational losses (e.g., lost production).

In addition, MI training and procedures heighten workforce efficiency and result in more consistent job performance. In general, personnel who are properly trained and have access to up-to-date, correct procedures (and other pertinent documentation) will perform tasks more consistently and efficiently. This improves downtime planning and eliminates many breakdowns.

With some effort, many of the benefits described in this section can be quantified by comparing equipment reliability performance (e.g., process availability, unplanned downtime) and workforce efficiency measures (e.g., repeat work, average repair times for tasks) before and after MI program implementation. Chapter 12 provides some suggested performance measures that can be used for such a comparison.

### 10.3.2 Cost Avoidance

Cost avoidance involves the return associated with avoiding the cost of an equipment failure. For example, an internal inspection of a storage tank that discovers that the tank bottom is thin helps a facility avoid the costs associated with the following issues:

- Further equipment damage, which could increase the repair cost
- Potential accident impacts (to personnel and/or related equipment) that could have resulted if a release had occurred
- Potential environmental impacts, including cleanup costs and adverse publicity, that could have resulted if a release had occurred
- Potential unscheduled downtime and the associated production losses

Measuring the monetary value of these avoided costs can be difficult. However, in the evolving field of cost-benefit analysis, some corporations are creating "equivalent pain" matrices (similar to risk matrices) and other tools designed for making just such measurements.

### 10.3.3 Regulatory Compliance and Industry Association Commitments

For many organizations, one objective of the MI program is compliance with relevant regulations and/or industry association commitments (e.g., commitment to comply with ACC's Responsible Care® initiative). The avoidance of regulatory costs, when quantified, is calculated based on (1) the magnitude of fines issued to date and (2) the harder-to-quantify benefits of industry and company reputation.

### 10.3.4 Reduced Liability and Reduced Damage to Corporate Reputation

MI programs focus on maintaining equipment integrity so that failures, especially catastrophic failures, do not occur. Therefore, an effective MI program will result in reduced risk of:

- Negative publicity and/or extended production outages that could result in lost market share
- Employee injuries that could result in litigation
- Offsite injuries and damage that could result in litigation
- Adverse public reaction/perception

In addition, an MI program that is effective in preventing equipment failures can have a positive impact on workforce morale and can promote good corporate citizenship.

## 10.4 REFERENCES

10-1. ABSG Consulting Inc., *Mechanical Integrity, Course 111*, Process Safety Institute, Houston, TX, 2004.

10-2. Kletz, T., *What Went Wrong: Case Histories of Process Plant Disasters*, 4th Edition, Elsevier Science & Technology Books, Burlington, MA, 1999.

10-3. Occupational Safety and Health Administration, *Process Safety Management of Highly Hazardous Chemicals*, 29 CFR Part 1910, Section 119, Washington, DC, 1992.

# 11
# RISK MANAGEMENT TOOLS

This chapter provides an overview of some engineering risk-based analytical techniques that can be used to assist in (1) making inspection, testing, and preventive maintenance (ITPM) task and frequency decisions and/or (2) advancing the ITPM decisions to a risk-based decision-making approach. Specifically, this chapter briefly discusses the application of the following techniques and tools to ITPM decisions:

- Failure modes and effects analysis (FMEA) and failure modes, effects, and criticality analysis (FMECA)
- Reliability-centered maintenance (RCM)
- Risk-based inspection (RBI)
- Layer of protection analysis (LOPA) and similar analysis approaches
- Fault tree and Markov analyses

Chapter 4 outlines an approach for determining equipment ITPM activities that is based primarily on codes, standards, recommended practices, and manufacturers' recommendations (i.e., recognized and generally accepted good engineering practices [RAGAGEPs]). Pertinent RAGAGEPs provide a prescriptive approach for defining ITPM tasks. However, when RAGAGEPs are not available, defining these tasks becomes more subjective. Some facilities desire a structured, performance-based approach for implementing ITPM activities over one based solely on prescriptive RAGAGEPs because these RAGAGEPs may not fully meet the facility's needs or ensure integrity of its equipment. Reasons for this include:

- ITPM tasks and their frequencies specified in the RAGAGEPs may not adequately address failures that may result from the actual design and operating conditions of the equipment in question. Because RAGAGEPs are often general (i.e., address the most common damage mechanisms and appropriate inspection practices), facilities may need to augment or enhance the requirements defined in the RAGAGEP.

- Requirements in the RAGAGEPs may be unwarranted and/or be too burdensome because of the operating conditions of the equipment.
- The facility may need to prioritize ITPM tasks in order to manage its resources, including the decision to allow select equipment to "run to failure."
- Pertinent RAGAGEPs may not be available for important facility equipment.

In recent years, facilities have applied risk-based analytical techniques in an effort to develop more performance-based mechanical integrity (MI) programs. In fact, some inspection standards, such as American Petroleum Institute (API) 510 and API 570, now include provisions for determining inspection requirements based on risk (References 11-1 and 11-2). Also, API and other organizations that publish RAGAGEPs have developed standards and recommended practices that encourage the use of risk-based techniques to define inspection and testing requirements, including:

- API Recommended Practice (RP) 580, *Risk-based Inspection*, and API Publ 581, *Base Resource Document – Risk-based Inspection*
- ANSI/Instrumentation, Systems, and Automation Society (ISA)-84.00.01-2004 Part 1 (International Electrotechnical Commission [IEC] 61511-1 Mod), *Functional Safety: Safety Instrumented Systems for the Process Industry Sector — Part 1: Framework, Definitions, System, Hardware and Software Requirements*

Many organizations find that using risk analysis techniques to determine ITPM tasks provides several benefits, such as:

- Assurance that a structured, systematic, and technically defensible approach is used to make decisions.
- Improved knowledge of the system operation and the cause/effect relationships that result from specific equipment failures.
- Confidence that MI resources are being focused on the most important failures by explicitly assessing the risk and then assigning ITPM resources to those areas for which they will be the most effective.

The next section summarizes some of the important attributes of the above-mentioned risk-based analytical techniques; the following sections of this chapter provide overviews of these techniques. The CD accompanying this book contains additional information and examples of the analysis approaches described in this chapter.

# 11. RISK MANAGEMENT TOOLS

## 11.1 INTRODUCTION TO COMMON RISK-BASED ANALYTICAL TECHNIQUES USED IN MI PROGRAMS

Each of these analytical techniques can assist decision making in the MI program; however, facilities need to recognize when these techniques are best applied, what decisions can be made with the results, and other issues related using the techniques (e.g., timing, advantages, resources). In general, the following are typical uses of these techniques in an MI program:

- FMEA/FMECA to identify and prioritize potential equipment failure modes that need to be addressed by ITPM tasks.
- RCM to optimize proactive maintenance tasks (e.g., predictive maintenance, preventive maintenance [PM], failure-finding tasks), usually applied to functional failures of active components (e.g., pumps failing off, erratic control).
- RBI to optimize inspection tasks and frequencies for fixed equipment (e.g., vessels, tanks, and piping) and pressure relief devices.
- LOPA and similar analysis approaches to define performance requirements for independent protection layers (IPLs), including safety instrumented functions (SIFs) (e.g., emergency shutdown [ESD] systems and functions).
- Fault tree and Markov analyses to verify that SIFs and IPLs designs meet the performance requirements (i.e., targeted probability of failure of demand).

Table 11-1 summarizes different key attributes for each of these techniques.

Sections 11.3 through 11.6 provide additional information on each of these analysis techniques. In addition, several texts/publications are available that provide more detailed information on these techniques. Some of these texts are:

- FMEA/FMECA: Center for Chemical Process Safety (CCPS), *Guidelines for Hazard Evaluation Procedures, Second Edition with Worked Examples*
- RCM: John Maubray, *Reliability-centered Maintenance and Reliability-centered Maintenance: RCMII*
- RBI: API RP 580, *Risk-based Inspection*, as a recommended practice, and API Publ 581, *Base Resource Document – Risk-based Inspection*
- LOPA: CCPS, *Layer of Protection Analysis, Simplified Process Risk Assessment*
- Fault tree and Markov analyses (as applied to SIF calculations): ISA technical reports TR84.00.02-2002, Parts 1 through 5, *Safety Instrumented Functions (SIF) — Safety Integrity Level (SIL) Evaluation Techniques*

Before discussing these techniques, the following section will describe basic risk concepts and how risk can be used in MI decisions.

## TABLE 11-1
## Summary of Risk-based Analytical Techniques

| | | | Protection Layer Analysis Techniques | |
|---|---|---|---|---|
| FMEA/FMECA | RCM | RBI | LOPA and Alternate Approaches | Fault Tree and Markov Analyses |

**Brief Description**

| FMEA/FMECA | RCM | RBI | LOPA and Alternate Approaches | Fault Tree and Markov Analyses |
|---|---|---|---|---|
| • Inductive reasoning approach that evaluates how the equipment can fail and the effect that these failures have on process or system performance, and ensures that appropriate safeguards against the failure(s) are in place.<br>• FMECA is an FMEA that assesses the criticality of the failure modes and resulting effects using qualitative, semi-quantitative, or quantitative risk measures. | Comprehensive review and analysis of systems and their components using (1) an FMEA/FMECA to identify potential equipment failures and their impact on system/process performance and (2) decision tree (or similar tools) to determine appropriate failure management strategies (e.g., ITPM tasks). | • Risk assessment and risk management process that assesses the likelihood and consequences of a loss of containment in process equipment used as an ongoing part of the MI program.<br>• It integrates the traditional RAGAGEP standards with flexibility to focus and optimize the activities on risk reduction by identifying higher risk equipment and failure mechanisms. | • LOPA - a semi-quantitative analysis of the risk of a scenario; each scenario has a consequence with its associated severity and one initiating event with its associated frequency; IPLs are evaluated for applicable risk reduction; additional layers, such as SIFs, can be added to meet a risk target.<br>• Alternate approaches – same analysis objectives as LOPA, but typically using more quantitative analysis approaches, such as event tree analysis. | Quantitative analysis tools used to estimate the unavailability (probability of failure on demand [PFD]) of protection layers, including SIFs. |

**Types of Equipment**

| FMEA/FMECA | RCM | RBI | LOPA and Alternate Approaches | Fault Tree and Markov Analyses |
|---|---|---|---|---|
| All types of equipment, but is typically best applied to mechanical (e.g., pumps, compressors) and electrical equipment. | Mechanical (e.g., pumps, compressors) and electrical equipment, and instrument systems. | Pressure vessels, storage tanks, and piping systems. In addition, recent use on pressure relief systems. | Process controls and safety interlocks, operator responses to alarms, relief devices, and other IPLs. | Process controls and safety interlocks, operator responses to alarms, relief devices, and other IPLs. |

**Suggested Application**

| FMEA/FMECA | RCM | RBI | LOPA and Alternate Approaches | Fault Tree and Markov Analyses |
|---|---|---|---|---|
| Critical and/or complex systems in which failure mode cause and/or effect on process/system performance is not known or well understood. | Critical and/or complex systems in which failure mode cause and/or effect on process/system performance is not known or well understood. | In-service inspection of pressure vessels, storage tanks, piping systems, and pressure relief devices for loss of containment issues. | Processes with higher risk accident scenarios (e.g., fire, explosions) that require more in-depth evaluation (than provided by process hazard analyses [PHAs]) to determine if sufficient layers of protection are provided to meet risk criteria. In addition, to determine the required PFD for safety systems, especially SIFs. | Evaluation/verification that SIF and IPL designs meet the PFD requirements (typically defined via LOPA or one of the alternate approaches). |

## 11. RISK MANAGEMENT TOOLS

### TABLE 11-1 (Continued)

| FMEA/FMECA | RCM | RBI | Protection Layer Analysis Techniques | |
|---|---|---|---|---|
| | | | LOPA and Alternate Approaches | Fault Tree and Markov Analyses |
| **Use of the Results in the ITPM Program** | | | | |
| • Identifies system/equipment failure modes and specific failure causes to be addressed by ITPM tasks.<br>• Provides risk/criticality rankings that can be used to establish ITPM task frequencies and to prioritize ITPM tasks. | • Identifies appropriate ITPM tasks and frequencies needed to address system/equipment failure modes and specific failure causes.<br>• Provides risk/criticality rankings that can be used to establish ITPM task frequencies and to prioritize ITPM tasks. | • Identifies inspection strategy for equipment (Note: Activities are based heavily on the API inspection standards, with added flexibility on the extent and frequency of inspections based on risk).<br>• Uses inspection results to update the extent and frequency of inspections in order to manage the likelihood of failure. | • Does not typically define ITPM tasks and frequencies directly or in detail.<br>• Can be used to identify critical safety systems needing to be addressed by ITPM tasks. | Defines the testing frequency required for SIFs (and potentially other IPLs) to achieve the required PFD. |
| **Use of the Results in Other MI Program Activities** | | | | |
| • Can be design-level quality assurance (QA) activity, as part of the design review to identify potential failures.<br>• Can also identify potential maintenance errors resulting in system/equipment failures that should be addressed by training or procedures.<br>• Can be used to develop an equipment troubleshooting guide. | • Can be design-level QA activity, as part of the design review to identify potential failures and develop strategies for managing the failures (e.g., redesign, start-up considerations).<br>• Can also identify potential maintenance errors resulting in system/equipment failures that should be addressed by training or procedures.<br>• Can be used to develop an equipment troubleshooting guide. | Typically, not applied to other MI program activities. | Used during the design phase to establish the risk reduction of IPLs, including the SIL of an SIF. | Used during the design phase to define specific design requirements (e.g., level of redundancy, dangerous failure rate targets) for SIFs. |

TABLE 11-1 (Continued)

| FMEA/FMECA | RCM | RBI | Protection Layer Analysis Techniques ||
| | | | LOPA and Alternate Approaches | Fault Tree and Markov Analyses |
|---|---|---|---|---|
| **Timing** | | | | |
| Initial development of the ITPM program when appropriate tasks are not apparent or results of an existing ITPM program are not adequate. | • Initial development of the ITPM program when appropriate tasks are not apparent or results of an existing ITPM program are not adequate.<br>• Initial design of system/equipment to identify opportunities to improve system/equipment reliability and integrity; this timing probably provides the greatest value, but is seldom done by facilities. | • Initial development of the ITPM program.<br>• Optimization of initial inspection efforts.<br>• Ongoing program to better focus and optimize regular inspections. | • Initial design of a process and its safety systems.<br>• Applicable in existing process units to check or verify the integrity of the existing systems. | • Initial design of SIFs and/or IPLs.<br>• Applicable in existing process units to check or verify SIF/IPL integrity of the existing systems. |
| **Effort/Resource Requirements** | | | | |
| Variable – can be as simple as using generic FMEA/FMECA results (i.e., templates) for standard equipment types, to more resource-intensive development of FMEA/FMECA results requiring input from reliability personnel, maintenance personnel, operations personnel, process engineers, and other engineering specialists if warranted. | Resource intensive; however can be reduced by the application of FMEA/FMECA templates and generic ITPM plan. These resource reduction efforts can decrease some RCM benefits (e.g., specific equipment failure management strategies). | Greater initial effort than a conventional program, but typically with rapid payback. | • LOPA requires a moderate level of resources; the analysis requires (or begins) a PHA to identify the accident scenarios, and then a team to evaluate the sufficiency of the IPLs.<br>• Alternate analysis approaches (e.g., event tree analysis) require more resources and can be resource intensive depending on the number and complexity of the IPLs, the availability and applicability of failure data, and other modeling needed to determine PFD of the IPLs. | • Moderate to high, depending on number and complexity of SIFs and IPLS evaluated, availability and applicability of failure data, and analysis technique used to model the PFD of SIFs/IPLs.<br>• ISA has developed simplified equations that can be used to determine the SIL/IPL PFD; these equations can reduce the resources required to determine SIFs/IPL PFDs. |

## 11. RISK MANAGEMENT TOOLS

**TABLE 11-1 (Continued)**

| FMEA/FMECA | RCM | RBI | Protection Layer Analysis Techniques | |
|---|---|---|---|---|
| | | | LOPA and Alternate Approaches | Fault Tree and Markov Analyses |

**Advantages**

| FMEA/FMECA | RCM | RBI | LOPA and Alternate Approaches | Fault Tree and Markov Analyses |
|---|---|---|---|---|
| • Thorough, logical approach for identifying and evaluating system/equipment failures and their importance (based on impact on system/process performance).<br>• Objective method for prioritizing equipment failures through the use of risk or criticality rankings. | • Thorough, logical approach for developing ITPM task plans (i.e., provides a rationale or basis for ITPM tasks for which no applicable RAGAGEPs are available).<br>• Effective in evaluating system/equipment designs and developing appropriate strategies for managing potential failures. | • Better focus on controlling risk instead of just generating inspection results.<br>• Lower costs.<br>• Less likely to lose sight of the high-risk items in the quantity of low-risk data. | • Compared to other techniques, LOPA is a simple approach for accomplishing the objective.<br>• Objectively evaluates the sufficiency of IPLs (i.e., less team judgment than more qualitative methods). | • More complex and rigorous approaches; both methods are good where common mode failures exist; Markov analysis is of particular value when a time dependency for detecting, repairing, or restoring a degraded system to full functionality exists.<br>• Objective evaluation of the sufficiency of IPLs (i.e., less team judgment than more qualitative methods). |

**Disadvantages**

| FMEA/FMECA | RCM | RBI | LOPA and Alternate Approaches | Fault Tree and Markov Analyses |
|---|---|---|---|---|
| • Can be resource intensive.<br>• Only identifies single initiated event failures.<br>• Results depend somewhat on analysis team's knowledge of the system/equipment. | • Tends to be resource intensive.<br>• Only identifies single initiated event failures.<br>• Results depend somewhat on analysis team's knowledge of the system/equipment. | • Requires greater technical knowledge/training to successfully set up and maintain than a conventional program. | • May yield overly conservative results for systems with potential common mode failures (e.g., interlocks within the same process control system). | • Most resource intensive.<br>• Requires highest level of training. |

## 11.2 INCORPORATING RISK INTO MI DECISIONS

A primary objective of an MI program is to reduce unreliable performance as a result of equipment failures (Reference 11-3). An MI program can accomplish this by identifying and prescribing tasks to prevent equipment failures, detect the onset of equipment failures, and/or discover hidden equipment failures before they impact system performance, safety, and/or the environment. Risk-based analytical techniques can be used to assess unreliable performance by identifying potential loss exposures and then estimating their risks. Based on this analysis, the facility staff can determine the most effective MI tasks for reducing those loss events.

Risk is the combination of the consequence and the frequency of an undesired event. Consequences may include safety, health, economic, environmental, or other types of potential losses. Likelihood addresses the frequency of a single occurrence of the undesired event (e.g., equipment failures) or the average frequency of occurrence of the undesired event. For example, a machine that fails twice per year causing $100,000 in lost production per failure event has the same risk as a machine that fails 10 times per year causing $20,000 in lost production per event. Over a year's time, the failure of each machine results in $200,000 in lost production. The economic risk associated with operating these two machines is equal. Using this example, the risk of operating the two machines is measured as an annual cost (dollars per year). This example expresses risk in terms of dollars to illustrate the calculation of risk. Other adverse effects (e.g., safety, environmental) can be measured with other units, such as injuries per year, affected area of fires per year, or releases of chemicals per year.

Understanding a system's risk can help a facility develop a performance-based ITPM strategy. Identifying higher risk areas provides better opportunities for risk mitigation. Priorities can be established by comparing systems, system functions, and component failure modes. Also, risk characterization tools can be used during MI task development to predict the effectiveness of the tasks being assigned. When assigning tasks, the objective is to mitigate the risk of a failure to an acceptable risk threshold, which can be considered as "balancing the risk." Risk is considered to be in the balanced state when the assigned maintenance tasks are effective in reducing the risk to an acceptable threshold.

An acceptable threshold is known as the risk acceptance criteria, the rate of loss that decision makers are willing to accept for a given consequence. This acceptable rate of loss will vary with many factors, including the type of loss, the magnitude of loss, and the individuals or groups affected by the loss. Risks that exceed the risk acceptance criteria require actions to lower the risk.

A common risk characterization tool used for MI decisions (and other risk decisions) is a risk matrix. A risk matrix is a matrix that has frequency on one axis and consequence on the other axis. Figure 11-1 is an example risk matrix with consequence on the horizontal axis and frequency on the vertical axis. The consequence and frequency ratings are typically defined as categories (i.e., range of consequences and frequency). These ratings can be defined qualitatively (e.g., insignificant, catastrophic, extremely remote, frequent) or quantitatively (e.g., 1 to 10 times per year, $1M to $10M). Each cell in the risk matrix represents

## 11. RISK MANAGEMENT TOOLS

a risk characterization that is defined by the intersection of the consequence and frequency ratings. In addition, risk acceptance criteria can also be included on the risk matrix by defining a risk level for each cell (e.g., high risk, medium-high risk, medium risk, and low risk). The results are groups of cells with the same risk level. These risk levels are then used to define the risk acceptance criteria by indicating which risk levels require reduction and which are acceptable. The risk acceptance criteria are often indicated by a line that divides the risk matrix into two regions: acceptable and unacceptable risk.

Another risk acceptance measure is known as "as low as reasonably practicable (ALARP)". An ALARP risk matrix typically contains three regions: intolerable risk, negligible risk, and ALARP. Figure 11-1 also presents an example risk matrix with these three regions. Risks that fall in the unacceptable region require actions to lower the risk into the ALARP or acceptable risk regions. Risks that fall in the acceptable risk region do not require any actions. Risks that fall in the ALARP region should be further evaluated to determine if the risk could be reduced by implementing reasonable risk-reduction actions. To use a risk matrix, the analysis team (1) selects a loss event of interest (e.g., fire/explosion caused by leak of flammable material), (2) determines the consequence and frequency categories that best characterize the event, (3) identifies the risk level on the matrix, and (4) bases the need for decisions on the risk level.

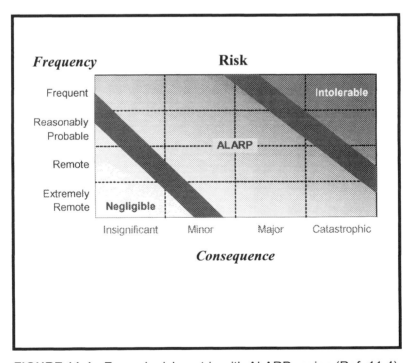

**FIGURE 11-1** Example risk matrix with ALARP region (Ref. 11-4).

## 11.3 FMEA/FMECA

Industry has used the FMEA technique to (1) identify potential equipment failures in complex systems, (2) understand the failure's effects on system performance (e.g., safety impacts, operational issues), (3) determine if sufficient safeguards are provided (e.g., equipment protection systems), and (4) identify equipment and system improvements. FMEAs have been used to evaluate many different types of systems, especially mechanical and electrical systems. FMEA is most commonly used in MI to analyze critical equipment and complex systems for which failure causes and/or impacts (effect) on process/system performance are not known or well understood.

FMEA is an inductive reasoning approach that (1) considers how the equipment can fail, (2) determines the effects that these failures have on process or system performance, and (3) ensures that appropriate safeguards against the failure(s) (including appropriate ITPM tasks) are in place. Another version of FMEA is FMECA. FMECA is an FMEA that assesses the criticality of the failure modes and resulting effects using qualitative, semi-quantitative, or quantitative risk measures.

FMEAs are often used during the design stage to identify and evaluate potential failures of mechanical and electrical systems. However, an FMEA can also be used on existing systems to provide a better understanding of potential failures, the impact of those failures, and the existing safeguards. These failure descriptions provide facility personnel with a basis for identifying design improvements and equipment failures requiring ITPM tasks. In addition, FMEAs can identify:

- Consequences of the equipment failures, allowing facility personnel to determine which equipment failures are most important to equipment integrity.
- Causes of the equipment failures so that ITPM tasks can be selected to address underlying causes (FMEA teams can also assess whether other means [e.g., operator training, better operating procedures] are more appropriate).
- Risk of the equipment failures so that equipment failures can be prioritized and the appropriate level of resources can be allotted.

An FMEA generally involves the following steps:

- Step 1 – Define the equipment and process of interest.
- Step 2 – Define the effects of interest to consider during the analysis.
- Step 3 – Subdivide the process into systems or equipment items for analysis.
- Step 4 – Identify potential failure modes of the systems/equipment items.
- Step 5 – Evaluate potential failure modes capable of producing effects of interest (i.e., list underlying causes and safeguards, evaluate risk, recommend changes needed).

# 11. RISK MANAGEMENT TOOLS

- Step 6 – Perform a quantitative evaluation (if needed or desired).

Table 11-2 provides an example worksheet from an FMEA (Reference 11-5). FMEAs are generally performed by a team that consists of personnel who are familiar with the design, operation, and maintenance of the system. A team leader with expertise in the analysis technique facilitates and documents the analysis. Additional information on FMEAs can be found in the CCPS book, *Guidelines for Hazard Evaluation Procedures, Second Edition with Worked Examples* (Reference 11-6).

FMEA and/or FMECA can be used in the ITPM task planning process to better understand the cause-effect relationship between equipment failures and resulting effects, usually in terms of loss events (e.g., safety events, environmental releases, production downtime). This attribute makes FMEA and/or FMECA a useful tool for the ITPM planning process. Benefits include:

- Identification of specific equipment failure modes/causes that ITPM tasks should focus on
- Documentation of the rationale for ITPM tasks decisions, especially run-to-failure decisions
- Risk ranking, which can be used to prioritize ITPM resources

Determining the appropriate level of analysis detail as early in the process as possible is one key to successfully using FMEA and/or FMECA to develop an ITPM task plan. An analysis performed with too little detail might not provide the desired information. On the other hand, too much detail will increase analysis time and effort with little benefit. Typically, the FMEA can be performed at the detail level that the ITPM tasks will be defined (e.g., pump, tank, vessel, piping circuit).

## 11.4 RCM

RCM analysis is an established technique that is used for defining maintenance tasks, including ITPM tasks. Specifically, RCM can be used in an MI program to evaluate critical and complex systems to determine: (1) which potential failures are important (i.e., highest risk) and (2) what ITPM tasks and task frequencies are needed to address the failure (i.e., best strategy for managing the failure). Originally developed for the aviation industry and later expanded to other industries, RCM provides a systematic approach for identifying potential failures that can affect process or system performance. The characteristics of the potential failures are then evaluated to determine (1) appropriate maintenance tasks or

## TABLE 11-2
### Sample FMEA Worksheet

| Item Number | Component | Failure Mode | Effects | Safeguards | Action |
|---|---|---|---|---|---|
| 1 | Temperature Control Valve (TCV-201) | a. Fails Open | Increased heating of the HCl column. Potential column overpressure and release of HCl | Multiple TIs on column | Add a high-pressure alarm |
| | | | | High-temperature alarm and interlock on column | Develop an emergency checklist for operators to follow on high temperatures |
| | | | | Excess overhead condensing capacity (spare condenser) | Function test valve operation and/or check positioner calibration |
| | | b. Fails Closed | HCl column cools down. No effect of interest | — | |
| | | c. Leaks Externally | Loss of steam. No effect of interest | — | |
| | | d. Leaks Internally | Increased heating of the HCl column; however, at a slow rate | Multiple TIs on column | |
| | | | Potential column overpressure and release of HCL | High-temperature alarm and interlock on column | |
| | | | | Excess overhead condensing capacity (spare condenser) | |
| | | | | Operators have ample time to diagnose problem and manually isolate TCV-201 | |

## 11. RISK MANAGEMENT TOOLS

(2) potential design or operational changes (referred to as one-time changes in RCM). The RCM approach is based on systematically answering the following seven questions (Reference 11-7):

1. What process/system functions need to be preserved? (For process safety, these functions will typically relate to containing hazardous materials or ensuring functionality of safety-critical or mitigation systems.)
2. How can the process/system fail to fulfill these functions (i.e., what functional failures can occur)?
3. What specific equipment failures can cause each functional failure?
4. What happens when the failure occurs (i.e., the effect on system performance)?
5. Why does the failure matter (i.e., the severity of the effect)?
6. What should be done to prevent or detect the failure (e.g., proactive maintenance, such as ITPM tasks)?
7. What should be done if appropriate maintenance cannot be found or is not effective (e.g., design or operational changes)?

An RCM analysis typically uses two analysis tools to answer these questions: FMEA and a decision tree. The FMEA is used to help answer questions 1 through 5, and the decision tree is used to answer questions 6 and 7. Table 11-3 provides an example FMEA from an RCM analysis, and Figure 11-2 provides an example RCM decision tree.

An RCM analysis generally includes the following steps:

- Step 1 – Define the system and its boundaries
- Step 2 – Define the system's functions and functional failures
- Step 3 – Conduct an FMEA
- Step 4 – Select failure management strategies (e.g., maintenance tasks, design or operational changes)
- Step 5 – Finalize the failure management strategy based on risk

A team is needed to perform the analysis. This team usually consists of personnel familiar with the design, operation, and maintenance of the system. A team leader with expertise in the RCM analysis technique facilitates and documents the analysis. Table 11-4 provides example results from an RCM analysis task selection activity.

Several textbooks have been written on RCM. Two of the more prominent textbooks are *Reliability-centered Maintenance* and *Reliability-centered Maintenance: RCMI*. In addition, the Society of Automotive Engineers has published an RCM Standard, JA1011, *Evaluation Criteria for Reliability-centered Maintenance Processes*.

## TABLE 11-3
### Sample RCM FMEA Worksheet

Equipment Item: Pump 1A, including the gearbox and the motor

| Failure Mode | Failure Characteristic | Hidden/ Evident | Effects | | | Risk Characterization[1] | | | | |
| --- | --- | --- | --- | --- | --- | --- | --- | --- | --- | --- |
| | | | Local | Functional Failure | End | C | UL | UR | ML | MR |
| External leak | Wear out | Evident | Release of hazardous material | Loss of containment | Potential severe injury to employees | Major | Occasional | Medium | Remote | Medium |
| Fails off | Random | Evident | Brief loss of flow until the spare pump is started | Transfer time too long | Brief interruption in production | Minor | Frequent | Medium | Occasional | Low |
| Degraded head | Wear out | Evident | Reduced flow of material | Transfer time too long | Production rate reduced | Moderate | Occasional | Medium | Remote | Low |

[1] *Risk characterization abbreviations:*
  *C is consequence (severity)*
  *UL is unmitigated likelihood*
  *UR is unmitigated risk*
  *ML is mitigated likelihood*
  *MR is mitigated risk*

# 11. RISK MANAGEMENT TOOLS

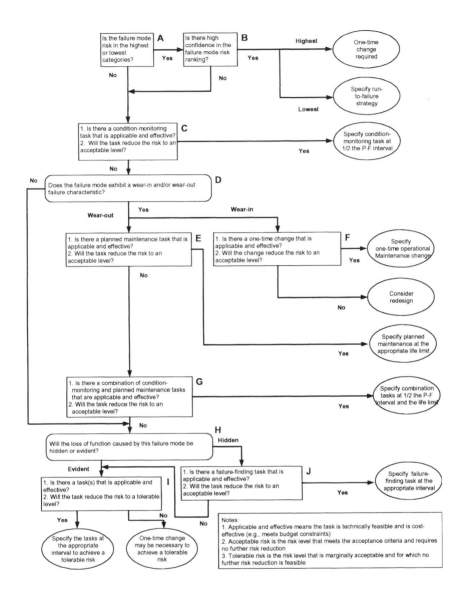

**FIGURE 11-2**  Example RCM decision tree.

**TABLE 11-4**
**Final Failure Management Strategy Selection for Pump 1A**

| Current Maintenance | | RCM-generated Maintenance | |
|---|---|---|---|
| Task | Interval | Task | Interval |
| Vibration analysis | Not used | Vibration analysis | 1 week |
| Rebuilding | 6 months | Rebuilding | 1 year |
| Lubrication | 3 months | Lubrication | 1 month |
| Visual inspection | 1 week | Performance monitoring | 1 week |

## 11.5 RISK-BASED INSPECTION

RBI is an inspection planning approach that is becoming more common in the chemical process industries (CPI) (Reference 11-7). RBI is a risk assessment and risk management tool that assesses the likelihood and consequence of a loss of containment in process equipment. It integrates the traditional RAGAGEP standards with flexibility to focus and optimize the activities on risk reduction by identifying higher risk equipment. RBI is typically used to develop and optimize the inspection plans for pressure vessels, storage tanks, piping, and relief devices. To assist in the development and implementation of RBI programs, API has published API RP 580, *Risk-based Inspection*, as a recommended practice and API Publ 581, *Base Resource Document – Risk-based Inspection*, as further guidance for the implementation of RBI.

Current RBI methodologies focus primarily on pressure vessels, storage tanks, and piping, and assess only the risks associated with losses of containment that can be affected by more or better inspections. This approach provides more focus on equipment with higher risks, through a systematic analysis of equipment damage mechanisms, condition, and the application of the most appropriate inspection techniques.

Conventional programs can be unfocused: applying the same, often less-effective inspections to virtually all equipment (e.g., pressure vessels, tanks, piping) at prescribed intervals. Too many facilities have broadly applied conventional inspection programs with excessive ultrasonic thickness (UT) measurements and have buried their personnel with measurement data, at the risk of losing sight of more critical equipment. A successful RBI program provides rule sets or strategies to guide appropriate inspections and to update inspection plans based on information from each round of inspections. Conventional programs sometimes provide little guidance for the inspector regarding when or why to increase or decrease inspection coverage or frequency. RBI programs can be much more responsive in updating inspection plans and can substantially reduce inspection and lost production cost while reducing risk. The relationship of risk level to inspection activity and inspection cost is illustrated in Figure 11-3.

## 11. RISK MANAGEMENT TOOLS

Focusing inspection activities on higher risk equipment provides more impact for the same effort and cost. As shown in Figure 11-3, some amount of equipment risk is not addressable through inspection; this risk may arise from design, operational, or maintenance issues. Also note the potential for risk to rise as typical conventional inspection programs become overloaded with data.

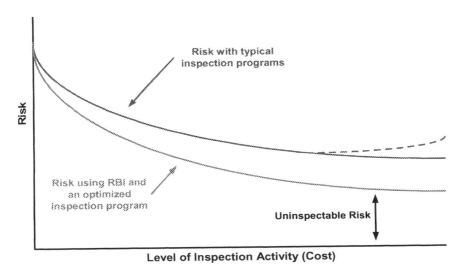

**FIGURE 11-3** Management of risk using RBI.

Because the implementation of an RBI program often involves extensive cost and time commitments, many facilities use project management procedures to guide the process. Expertise in RBI methodology and implementation will be needed. Whether such expertise is developed in-house or an RBI consultant is used, facilities need to provide work processes, procedures, inspection strategies, training, and software tools. Facility resources from inspection, process engineering, operations, and maintenance groups will also need to be tapped to provide input and expertise on the process and its equipment. The facility's corrosion engineers, equipment engineers, and risk analysts should be involved, as well. Features of RBI that differ from conventional MI programs are:

- Equipment and process data. Data needs are similar to ITPM planning data needs discussed in Section 4.1.3, but are even more fundamental to the analysis. Necessary equipment data include design temperatures, design pressures, materials of construction, size, details on stress relief, insulation, coatings/linings, current condition, and number and type of prior inspections. Also, the following process information is needed: process stream compositions, fluid properties, operating temperatures, operating pressures, flammability, toxicity, and inventories. The analysis will focus on the normal/routine process stream compositions; however,

nonroutine operating modes (e.g., startup, shutdown) need to be considered, as do abnormal conditions if they can initiate or accelerate damage to the equipment.
- Risk modeling. Personnel need to identify the underlying methodology for determining the likelihood and consequences of a release. Facilities typically use computer software to support the methodology; to properly interpret the results, personnel must understand the methods and assumptions used.
- Inspection strategies. RBI programs use a set of rules or guidelines for determining appropriate inspection methods, levels of inspection, and maximum intervals. These guidelines provide the methodology for creating inspection plans for each piece of equipment, based on its risk ranking, equipment type, and deterioration mechanisms.
- Inspection planning. RBI takes current inspection results, enters data into the risk model, recalculates the risk ranking, and modifies the inspection plan accordingly, based on the inspection strategy and the inspector's expertise.
- Management systems and tools. RBI generally uses work processes and computer tools to collect, interpret, integrate, and report the inspection data, as well as to plan and schedule inspection tasks. Management of the RBI program also involves reporting on activities, status, exceptions, and trends.

Figure 11-4 is a flowchart of an RBI program. Note that, at the front end of the work process and at the update cycle, RBI requires activities that are distinctly different from those of traditional MI approaches. The shaded boxes, and the arrows entering and exiting those boxes, represent these activities. The following paragraphs provide an overview of these front-end processes.

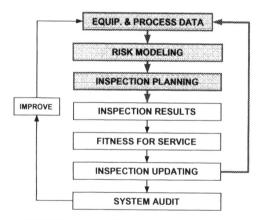

**FIGURE 11-4** RBI program flowchart.

## 11. RISK MANAGEMENT TOOLS

### 11.5.1 Equipment and Process Data

RBI begins with a formal "corrosion study" to identify potential damage mechanisms and rates. Damage mechanisms can be any combination of corrosion, cracking, or other degradation types. Because inspections focus on the damage mechanisms identified, failure to identify a potential damage mechanism could result in very significant equipment damage. Also, corrosion rates can significantly influence the risk ranking's estimation of failure likelihood and determination of inspection intervals.

The corrosion study should involve all appropriate personnel (e.g., inspection, process operations, maintenance, metallurgical, and corrosion engineers), and these personnel should have access to all available information. The main focus of the study is on the normal/routine process stream conditions; however, nonroutine (e.g., startup, shutdown) and some abnormal conditions should be considered if they can initiate or accelerate damage in the equipment. For example, if once a year on average, acid contaminates a system or a cooler is fouled and the process temperatures get high, the upsets will have a significant impact on the condition of the equipment. In certain cases, setting critical operating parameters for the equipment may be advisable; procedures should include steps for notifying the Inspection department if these parameters are exceeded. The current condition and past service of the equipment should be reviewed in this study to ensure that the written history is complete. The corrosion study provides the basis for the risk ranking, and the results should be documented for future reference.

### 11.5.2 Risk Modeling

Facilities use risk modeling in RBI to prioritize or rank the equipment for inspection. A high level of precision is not necessary to accomplish this ranking. Risk will be assessed for every piece of equipment and every piping circuit in the program, and the risk will be recalculated following every inspection; therefore, the assessment must be quick and easy. Organizations usually need an automated system that can be run by a trained inspector or inspection engineer as part of the normal work process.

The API RBI methodology describes several levels of analysis: quantitative, semi-quantitative, and qualitative. Different programs use different combinations, but most are mixes of numerical inputs and factors or groupings (i.e., semi-quantitative).

Detailed review of consequence modeling is beyond the scope of this book. Consequence modeling usually includes a simplified dispersion modeling of leakage and the area affected by a toxic release, fire, or explosion. Business and some environmental consequence data may be available, but such data may be limited to a cost-per-event input. In general, facilities determine consequences by evaluating the properties of the process stream, the size (or series of sizes) of the leak, and the duration of the release. Some questions to consider:

- What chemicals are not already dispersion modeled?
- How will the variety of chemical mixtures be handled?

- How are multiple consequence types (i.e., flammable and toxic) for the same equipment handled?
- How is leak duration determined?
- What are the program simplifications, default values, or assumptions that affect the modeling? An example might be, how are leak detection and isolation times determined?

Company decision makers must understand and be comfortable with the assumptions and consequence modeling issues at the start of the RBI program implementation. Changing the risk ranking inputs after the inspection planning has been completed can be costly because facility personnel must then revise the inspection plans.

Likelihood (frequency) of failure modeling is based on many factors. Inputs to the likelihood evaluation include generic equipment failure rates for equipment by type and for piping by size. Some evaluation of initial design robustness (over-design, safety factors, complexity, etc.) is included. API provides technical modules for different degradation mechanisms. The primary inputs to these modules include the rate of damage and the effectiveness of past inspections in forecasting condition up until the next inspection. Determining the likelihood of failure is greatly influenced by a facility's effectiveness or confidence in knowing the current equipment condition and the actual rates of damage. Details on this information are available from the API *Base Resource Document – Risk-based Inspection*.

After the initial risk rankings are determined, the results should be reviewed, both as a check for input errors and to define follow-up inspections. For example, because of a lack of inspection data, the corrosion study could have resulted in a conservatively estimated pressure vessel corrosion rate of 0.005 inches/year, yet the risk ranking reveals an extremely high likelihood of failure for the vessel. Why is this? Perhaps the vessel is 40 years old (40 years x 0.005 inches/year = 0.200 inches of corrosion loss) and, based on the input corrosion rate, almost no vessel wall remains. A few UT measures could quickly confirm that a corrosion rate of 0.001 to 0.002 inches/year is more reasonable and that imminent failure is not a risk.

### 11.5.3 Inspection Planning Strategies/Guidelines

For each equipment type and damage mechanism, organizations will need a strategy or guideline covering the types of inspections to perform, the extent of each inspection, and the interval for reinspection. An inspection guideline provides uniformity in the application of inspection tasks and may simplify the training for new inspectors. The inspection coverage and/or inspection intervals may vary with the risk ranking of the equipment. If an organization decides to use a contractor's guidelines, management should review and modify the guidelines prior to starting the inspection plans for specific equipment to ensure that the contractor's guidelines are consistent with the company's practices and policies, as

## 11. RISK MANAGEMENT TOOLS

well as compliant with jurisdictional requirements. Table 11-5 is a simple example inspection strategy.

**TABLE 11-5**
**Example Inspection Strategy for Pressure Vessel - External Deterioration Identified by External Visual Inspection**

|  | Risk Rank | | | |
| --- | --- | --- | --- | --- |
|  | Low | Medium | Medium-High | High |
| Minimum External Visual Coverage [1] | 100% External Visual | 100% External Visual | 100% External Visual | 100% External Visual |
| Maximum Interval (yr), or ½ Remaining Life | 15 | 10 | 10 | 5 |
| Maximum Interval Allowed by Jurisdiction | Depends on Jurisdictional Requirements | Depends on Jurisdictional Requirements | Depends on Jurisdictional Requirements | Depends on Jurisdictional Requirements |
| Inspection Confidence:<br>• Noninsulated<br>• Insulated | Very High<br>Medium | Very High<br>Medium | Very High<br>Medium | Very High<br>Medium |

[1] *Formal external visual inspections are performed on pressure vessels to help prevent failures by identifying those components that are deteriorating externally. External deterioration may include external corrosion, corrosion under insulation, fatigue, factors that may contribute to flange leaks, or indications of other failure modes.*

In Table 11-5, the inspection interval varies with the risk ranking, but is limited by the API Pressure Vessel Inspection Code to half the projected remaining life. Jurisdictional requirements, if any, can be listed. The inspection confidence for this visual inspection is used for some decisions. For example, equipment with a high risk rank that is subject to corrosion under insulation may, at the inspector's discretion, require additional inspection. In addition, an inspection strategy might adjust the inspection coverage or require more sophisticated inspection methods (with varying degrees of coverage) for different risk rankings.

### 11.5.4 Other RBI Program Issues

When implementing an RBI program at a facility, several additional issues must be appropriately managed by facility personnel for the RBI program to be successful. The following sections briefly discuss some of these issues.

*Management Systems and Software.* The amount of equipment data handled, the rate of change in the inspection task plans, changes to the inspection interval, and the complexity of the work process all increase with RBI. (Note that the volume of inspection data should decrease.) Coordinating different software tools for risk modeling, inspection planning and scheduling, and inspection data collection and analysis, can be a challenge; an organization should consider using integrated packages to simplify management of the system and increase efficiency. Also, ensure that the system can readily run reports and queries. The information retrieved may include summaries of inspections, scheduling information for upcoming and overdue inspections, and other exception reporting. Programs that feature different levels of authority and broad access to the data are desirable. Section 10.2 contains additional information on software used to manage inspection data and RBI studies.

*Management of Change (MOC) Feedback.* To keep the RBI program in line with the actual plant operating conditions, program updates are needed when process and operational changes and equipment changes (e.g., replacements, major repairs, materials of construction changes) occur. A facility's MOC work process may need to be revised to ensure that the RBI program receives this information.

*Operation Outside of Critical Operating Parameters.* Facilities may need to implement a procedure or system for reporting the kinds of upset conditions that can initiate or accelerate equipment damage (if these types of events were identified in the corrosion study) to ensure that the potential impacts of these events are assessed. Understanding and communicating the degree of excursion and the duration are necessary for making appropriate changes to the inspection plan.

*Integration of the Inspection Schedule with Turnaround Planning.* The inspection schedules and turnaround plans are interdependent: inspection requirements frequently drive turnaround activities and timing. Some external or online inspections should be performed well in advance of the turnaround so that the inspection results are available to those performing the turnaround work.

*Managing Schedule Extensions.* "The internal inspection is due this month but we just received a new product order. Is it okay to postpone the inspection for … months?" Many facilities routinely address this type of question. RBI provides organizations a new tool to help evaluate these cases. The risk model usually includes degradation rate confidence measures; facility personnel can use "what-if" cases with a longer interval to begin to understand the impact of the schedule extension on risk.

*Jurisdictional Issues.* Some state and city pressure vessel regulations may recognize RBI. Facilities should check with local jurisdictions to determine the acceptability of RBI. Regulators (and insurers) have generally been supportive of RBI, but they have voiced some concerns that improperly designed programs are focusing on cost savings rather than risk reduction. For some equipment, regulations may not permit reductions in inspection efforts.

*Acceptable Risk.* An RBI program should help reduce the likelihood of failure of higher-risk equipment, provided it is "inspectable risk" (i.e., risk that can be mitigated by increased inspection activity). The program could also provide leeway for some lower-risk events to increase in likelihood, as long as the risk ranking does not rise too much. The inspection strategies or guidelines should control these risk fluctuations; however, facilities should track and monitor RBI performance by periodically checking the population of each risk level (i.e., each box on a risk matrix) and noting trends. Periodic review also helps to identify higher-risk candidates for design, process, or operational changes when inspection is not effective at reducing the risk level.

## 11.6 PROTECTION LAYER ANALYSIS TECHNIQUES

As defined in American National Standards Institute (ANSI)/ISA-84.00.01-2004 Part 1 (IEC 61511-1 Mod), *Functional Safety: Safety Instrumented Systems for the Process Industry Secto — Part 1: Framework, Definitions, System, Hardware and Software Requirements*, an SIF is "a system composed of sensors, logic solvers, and final control elements for the purpose of taking the process to a safe state when predetermined conditions are violated." (Other terms used for SIF include ESD system, safety shutdown system, and safety interlock system.) (Note: A safety instrumented system [SIS] is a collection of SIFs on the same platform.) ISA S84.00.01 also defines a safety life cycle that includes the application of two analyses:

1. LOPA or an alternate tool, to determine if additional IPLs, including an SIF, are needed to achieve acceptable risk levels and, if acceptable risk is not achieved, to define the required risk reduction. Alternate tools include event tree analysis.
2. Reliability/unavailability analysis (e.g., component failure rate, redundancy) to determine the protection layer (including SIF) configuration and testing frequency needed to achieve the required risk reduction. Tools include fault tree analysis (FTA), simplified equations, and Markov analysis.

The results of these two analyses are the performance criteria (in terms of the probability of failure on demand [PFD]) for each SIF, which dictates much of the SIF design (e.g., acceptable equipment failure rate, level of redundancy) and defines the SIF's maintenance requirements, including the basics of ITPM tasks and the task frequencies. The following paragraphs provide an overview of these two types of analyses.

LOPA is a simplified risk analysis approach that can be used to (1) identify the need for additional IPLs, including an SIF, and (2) determine the performance required by each protection layer to achieve an acceptable level of risk. LOPA is an analysis tool that builds on the information developed during a hazard study (e.g., a PHA). Specifically, the hazard study is used to identify (1) the accident scenarios to be evaluated and (2) the non-SIF safeguards that are in place. (Note: Non-SIF safeguards are safeguards that are not part of a defined SIS, such as relief devices, basic process controls, alarms with operator response, mechanical safety devices, and containment dikes.) LOPA is performed using order-of-magnitude estimates of the cause frequency, consequence severity, and likelihood of failure of the non-SIF safeguards to determine if an SIF is needed to achieve an acceptable risk level. Qualitative rules (e.g., a risk matrix) or a combination of approaches can also be used to perform a LOPA. A LOPA generally includes the following steps (Reference 11-8):

- Step 1 – Identify accident scenarios (typically, a PHA is used to identify the scenarios).
- Step 2 – Identify the initiating event(s) for each scenario and estimate the initiating event frequency.
- Step 3 – Identify the consequence for each scenario and estimate its severity.
- Step 4 – Identify the IPLs for each scenario and estimate the likelihood of failure for each IPL (i.e., the PFD).
- Step 5 – Calculate the risk of the scenario by combining the initiating event frequency, the PFD for IPLs, and the consequence severity.
- Step 6 – Evaluate the risk to determine if an SIF is needed and/or to determine the performance required (e.g., PFD needed by the SIF[s]) to achieve an acceptable level of risk.

Figure 11-5 provides an example of a completed LOPA worksheet (Reference 11-9). LOPA can be performed as part of a PHA; however, most organizations use a separate team to perform the LOPA. The LOPA results can be used in the ITPM task planning process to:

- Identify the important non-SIF safeguards that need to be maintained by the MI program
- Define the performance requirements for the SIFs, including the type and frequency of testing needed to ensure that the SIF achieves the desired performance

Additional details on LOPA are provided in the CCPS book, *Layer of Protection Analysis, Simplified Process Risk Assessment*.

The results of a LOPA are then used to design the SIF (e.g., identify the redundancy needed, control system architecture, component failure rate). Specifically, the LOPA identifies (1) the safety function (i.e., what the SIF has to do) and (2) the required risk reduction (e.g., PFD required). The risk reduction is defined in terms of an SIL. ISA S84.00.01 defines three SIL categories, as shown in Table 11-6 (Reference 11-10).

## 11. RISK MANAGEMENT TOOLS 227

| Layer of Protection Analysis Summary Sheet: Current Situation (as-is) | | | | Side 1 | |
|---|---|---|---|---|---|
| Estimated By: | Joe Miner (UMR 1969) | Date Estimated: | 9/1/2004 | | Reactor 2700 - AMD Batch Reactor |
| Scenario Name: | Water addition to reactor R2700 via Nitrogen system | Related Deviations: | Reverse flow of water into nitrogen system from tank inerting systems | | |
| Initiating Event: | Water in Nitrogen System due to contamination via reverse flow from process equipment | Enabling Event/Condition Frequency: 1.0E+00 | | | |
| | | Initiating Event Frequency (accounting for enabling): 1.0E+00 | | | |
| Ultimate, Unmitigated Consequence: | Reaction of water with reactor contents leading to a runaway reaction, gas evolution and potential catastrophic failure of 2700. | Unmitigated Consequence Category: 2 | | | |
| Unmitigated Risk: | Risk level is 2 - Unacceptable risk | Initial Placement on Risk Matrix: Consequence 2 / Likelihood 1 | | Risk ($/yr): | |
| | Safeguards | Credits for Each Safeguards: | | Description of Safeguards | |
| | Reverse flow check valves on all nitrogen inerting systems | 1.0E-01 | | Mechanical Device | |
| | Relief valve on R2700 sized to protect reactor when 200 gallons | 1.0E-01 | | Mechanical device in fouling service, no history of pluggage | |
| | Composite Unavailability of Safeguards: | 1.0E-02 | | | |
| | Mitigated Frequency of Event/Consequence As-is: | 1.0E-02 | | | |
| | Mitigated Consequence As-is: No change to consequences | Mitigated Consequence Category: 2 | | | |
| | Mitigated Risk As-is: Risk category is 3 - Unacceptable risk - Recommendation required | As-is Placement on Risk Matrix: Consequence 2 / Likelihood 3 | | Risk ($/yr): | |
| NOTE: If Mitigated Risk As-is is Unacceptable, Proceed to Recommendation Evaluation on Back | | | | | |

**FIGURE 11-5** Sample LOPA worksheet (Page 1 of 2).

The SIL and $PFD_{AVG}$ are used to design the SIF. The SIL requires specific test procedures and frequencies to avoid systematic error and to provide required fault tolerance. As part of the design, the designers typically perform an unavailability analysis to evaluate if a proposed design configuration and testing frequency will achieve the required average PFD (i.e., SIL). This analysis is typically performed using simplified equations, fault trees, or Markov models. In addition, ISA has compiled simplified reliability equations that can be used to perform the analysis. ISA technical reports TR84.00.02-2002, Parts 1 through 5, *Safety Instrumented Functions (SIF) – Safety Integrity Level (SIL) Evaluation Techniques*, provide detailed information on the use of these analysis techniques to calculate the average PFD for proposed SIF designs.

The unavailability analysis is used to determine the testing frequency for each SIF and the type of testing required (e.g., calibration, functional testing). Also, SIS designers are to ensure that the SIF designs include design features (e.g., bypass valves) that will allow the SIF to be tested as required. ISA S84.01, *Application of Safety Instrumented Systems for the Process Industries*, also includes requirements for installation, commissioning, pre-startup acceptance testing, operations, MOC, and decommissioning (Reference 11-11). These requirements essentially outline a quality assurance system for SISs.

## Layer of Protection Analysis Summary Sheet: Recommendation Evaluation — Side 2

**Reason(s) Why Current Mitigation is Not Sufficient:**
Current risk falls into the unacceptable risk matrix

| Recommendations: | Credit for Each Recommendation: | Description |
|---|---|---|
| Water detection and shutdown system on nitrogen system. Install a small pot in nitrogen system with level detection sensors and interlock to stop flow of nitrogen to R2700 upon detection of liquid in pot. | 1.00E-02 | Install as SIL 2 interlock per ANSI/ISA S84.01 - 1996 |
| Revise maintenance and testing procedure of relief valve on R2700 to reduce potential for pluggage. | 1.00E+00 | Procedural Safeguard - No credit allowed until improvement documented for at least 2 years. |

| Revised List of Pertinent Safeguards: | Credits for Each Safeguards: | Description of Safeguard: |
|---|---|---|
| Water detection and shutdown system on nitrogen system. Install a small pot in nitrogen system with level detection sensors and interlock to stop flow of nitrogen to R2700 upon detection of liquid in pot. | 1.0E-02 | Install as SIL 2 interlock per ANSI/ISA S84.01 - 1996 |
| Reverse flow check valves on all nitrogen inerting systems | 1.0E-01 | Existing Mechanical Device |
| Relief valve on R2700 sized to protect reactor when 200 gallons water | 1.0E-01 | Existing Mechanical Device |

**Composite Unavailability of Safeguards(revised):** 1.0E-04

**Mitigated Frequency of Event/Consequence (revised):** 1.0E-04

**Mitigated Consequence (revised):**
No change on consequence - If water enters reactor, explosion may occur

**Mitigated Consequence Category (revised):** 2

**Revised Risk Rank:** 8 — Acceptable risk

**Revised Placement on Risk Matrix:** Consequence 2 | Likelihood 5

**Risk ($/yr):**

LOPA Form 1, Revised 1Sept04

**FIGURE 11-5** (Page 2 of 2)

**TABLE 11-6**
**ISA S84.01 SILs**

| SIL | Average Probability of Failure on Demand ($PFD_{AVG}$) |
|---|---|
| 1 | $10^{-1}$ to $10^{-2}$ |
| 2 | $10^{-2}$ to $10^{-3}$ |
| 3 | $10^{-3}$ to $10^{-4}$ |

## 11.7 REFERENCES

11-1 American Petroleum Institute, *Pressure Vessel Inspection Code: Maintenance Inspection, Rating, Repair and Alteration*, API 510, Washington, DC, 2003.

11-2 American Petroleum Institute, *Piping Inspection Code: Inspection, Repair, Alteration, and Re-rating of In-service Piping*, API 570, Washington, DC, 2003.

11-3 Montgomery, R. and W. Satterfield, *Applications of Risk-based Decision-making Tools for Process Equipment Maintenance*, presented at ASME Pressure Vessel and Piping Conference, Cleveland, OH, 2002.

11-4 Trinker, D., *A Survey of Risk Tolerance Criteria for Acute Process Related Industrial Hazards*, ABSG Consulting Inc., Knoxville, TN, 2000.

11-5 ABSG Consulting Inc., *Advanced Process Hazard Analysis Leader Techniques, Course 104*, Process Safety Institute, Houston, TX, 2004.

11-6 American Institute of Chemical Engineers, *Guidelines for Hazard Evaluation Procedures, Second Edition with Worked Examples*, Center for Chemical Process Safety, New York, NY, 1992.

11-7 SAE International, *Evaluation Criteria for Reliability-Centered Maintenance Processes*, SAE Standard JA1011, Washington, DC, 1999.

11-8 Folk, T., *Risk Based Approach to Mechanical Integrity Success in Implementation*, American Institute of Chemical Engineers, Spring National Meeting Process Plant Safety Symposium, New Orleans, LA, 2003.

11-9 American Institute of Chemical Engineers, *Layer of Protection Analysis, Simplified Process Risk Assessment*, Center for Chemical Process Safety, New York, NY, 2001.

11-10 ABSG Consulting Inc., *Layer of Protection Analysis, Course 209*, Process Safety Institute, Houston, TX, 2004.

11-11 The International Society for Measurement and Control, *Functional Safety: Safety Instrumented Systems for the Process Industry Secto — Part 1: Framework, Definitions, System, Hardware and Software Requirements*, ANSI/ISA-84.00.01-2004 Part 1 (IEC 61511-1 Mod), Research Triangle Park, NC, 2004.

### Additional Sources

American Petroleum Institute, *Risk-Based Inspection*, API RP 580, Washington, DC, 2002.

American Petroleum Institute, *Base Resource Document – Risk-based Inspection*, API Publ 581, Washington, DC, 2000.

Moubray, J., *Reliability Centered Maintenance: RCMII*, Industrial Press Inc., New York, NY, 1997.

Smith, A., *Reliability Centered Maintenance*, McGraw-Hill Inc., New York, NY, 1993.

International Maritime Organization, *Guidelines for Formal Safety Assessment (FSA) for Use in the IMO Rule Making Process*, MSC/Circ. 1023–MEPC/Circ. 392, London, England, 2002.

The International Society for Measurement and Control, *Safety Instrumented Functions (SIF) — Safety Integrity Level (SIL) Evaluation Techniques, Part 1: Introduction*, ISA-TR84.00.02-2002, Research Triangle Park, NC, 2002.

The International Society for Measurement and Control, *Safety Instrumented Functions (SIF) — Safety Integrity Level (SIL) Evaluation Techniques, Part 2: Determining the SIL of a SIF via Simplified Equations*, ISA-TR84.00.02-2002, Research Triangle Park, NC, 2002.

The International Society for Measurement and Control, *Safety Instrumented Functions (SIF) — Safety Integrity Level (SIL) Evaluation Techniques, Part 3: Determining the SIL of a SIF via Fault Tree Analysis*, ISA-TR84.00.02-2002, Research Triangle Park, NC, 2002.

The International Society for Measurement and Control, *Safety Instrumented Functions (SIF) — Safety Integrity Level (SIL) Evaluation Techniques, Part 4: Determining the SIL of a SIF via Markov Analysis*, ISA-TR84.00.02-2002, Research Triangle Park, NC, 2002.

The International Society for Measurement and Control, *Safety Instrumented Functions (SIF) — Safety Integrity Level (SIL) Evaluation Techniques, Part 5: Determining the PFD of Logic Solvers via Markov Analysis*, ISA-TR84.00.02-2002, Research Triangle Park, NC, 2002.

# 12
# CONTINUOUS IMPROVEMENT OF MI PROGRAMS

While much of a facility's mechanical integrity (MI) effort is generally directed at developing and implementing an MI program, successful facilities also have an objective and desire to continuously improve the program. Continuous improvement efforts often take the form of the following activities.

***Conducting Periodic Audits of Program Activities.*** Audits are typically used to evaluate how the MI program's management systems (e.g., quality assurance [QA] and inspection, testing, and preventive maintenance [ITPM] programs) are functioning and how they satisfy relevant requirements (e.g., process safety regulations). In fact, process safety regulations require periodic audits of the MI program (and other process safety management [PSM] systems) (Reference 12-1).

***Establishing Performance Measurement Systems.*** Effective performance measures can help organizations evaluate the ongoing performance of the overall MI program and key MI program activities (e.g., compliance with the ITPM task schedule). The measures should include both direct measurement of MI program performance and leading indicators of MI program performance (i.e., predictors of the program's performance).

***Learning Lessons from Equipment Failures.*** Another improvement activity is to systematically evaluate equipment failures, including near misses, using structured equipment failure and root cause analysis (RCA) processes. Using systematic analyses provides an effective means to identify the contributing and root causes of failures. From the results of these processes, personnel can make recommendations for eliminating these causes, and management can develop and implement appropriate actions to eliminate or reduce the likelihood of the failure. In addition, lessons learned from these analyses should be shared with others who could benefit, both within your facility and at other facilities.

These activities are included in Figure 12-1 (Reference 12-2), which shows a simple model that illustrates how continuous improvement efforts can contribute to the overall performance objectives of the MI program. Facilities should define a procedure for ensuring that the results from these continuous improvement

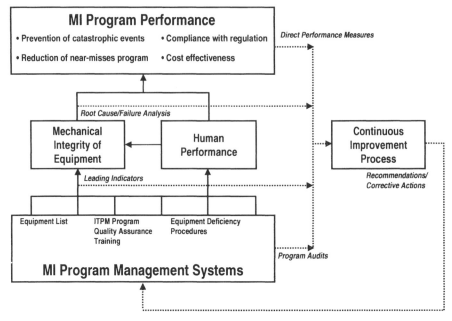

**FIGURE 12-1**  MI program continuous improvement model.

activities are implemented and evaluated for effectiveness. Continuous improvement activities can be most effective if (1) they produce corrective actions that are practical (e.g., technically feasible, have technical merit) and cost-effective and (2) these corrective actions are properly implemented. Therefore, each continuous improvement activity should include steps for:

- Conducting a management-level review of recommendations to ensure that the recommendations are effective and practical
- Communicating rejected recommendations (and the basis for the rejection) to the appropriate teams
- Generating appropriate corrective actions based on a review of the recommendations and/or other recommendations resulting from the management review
- Reviewing and approving corrective actions by management
- Prioritizing corrective actions for implementation, including identifying which corrective actions must be in place before critical process activities occur (such as placing deficient equipment in service or restarting a process)
- Periodic tracking (e.g., monthly, quarterly) of the implementation of corrective actions by assigned individuals and management
- Following up to ensure that the corrective actions have been appropriately implemented by assigned individuals and management

# 12. CONTINUOUS IMPROVEMENT OF MI PROGRAMS

The remaining sections of this chapter provide an overview of:

- Program audits
- Performance measurement and monitoring
- Equipment failure analysis and RCAs

Additional information on these topics can be found in the following Center for Chemical Process Safety (CCPS) publications:

- *Guidelines for Auditing Process Safety Management Systems*
- *If You Can't Measure It, You Can't Control It: ProSmart® Process Safety Management*
- *Guidelines for Investigating Chemical Process Incidents* (2nd Edition)

## 12.2 PROGRAM AUDITS

Program auditing is one of the basic tools for continuous improvement of an MI program. A good audit is a proactive means to uncover program deficiencies that can be corrected before an incident occurs.

Various aspects of the MI program can be readily audited. Some audits target select areas of suspected weakness, while others look at the overall program. Some of the more common program features to examine are:

- MI program equipment. Is all equipment whose failure could result in consequences of interest included in the MI program? Is all equipment used to prevent hazardous/unwanted conditions, or to detect and mitigate loss events, included in the MI program?
- ITPM tasks and schedules. Have ITPM and anticipated repair/rebuild activities been appropriately determined and their frequencies established? Are the ITPM tasks and schedules consistent with applicable recognized and generally accepted good engineering practices (RAGAGEPs)? Are ITPM tasks performed thoroughly and on schedule? Are ITPM tasks properly documented?
- Written procedures for MI program tasks. Do written procedures exist for all MI program tasks? Do the procedures contain adequate information, and are they kept current? Is work performed in accordance with the procedures? Has the facility implemented a means or program (e.g., periodic audits by supervision) to ensure that procedures are accurate and personnel are performing tasks in accordance with the procedures?
- MI training. Are personnel sufficiently trained in an overview of the process and its associated hazards, and in the procedures applicable to their job tasks, to ensure that the tasks can be performed correctly, safely, and consistently? Do craftspersons and inspectors have needed qualifications/ certifications?

- Correcting equipment deficiencies. Is a program/process in place to ensure that equipment deficiencies are identified and communicated? Are ITPM results reviewed for potential deficiencies? Are continued operation and associated temporary corrective measures (e.g., temporary repairs) reviewed and approved by appropriate personnel? Are temporary corrective actions communicated? Are temporary actions tracked to ensure that they are corrected in a timely manner?
- QA for equipment fabrication and installation and spare parts/equipment. Is a program/process in place to ensure that equipment is fabricated and installed in accordance with specifications? Are materials of construction verified for equipment/spare parts and/or maintenance materials? Has a contractor selection and auditing process been established? Are off-the-shelf spare parts, pieces of equipment, and other materials/supplies (e.g., bolts, fittings, gaskets, o-rings, welding rods) correctly ordered, received, placed into inventory, and removed from inventory before use or installation?

Comparing written records and policies with field observations and personnel interviews provides an auditor with a more accurate picture of the complete MI program. More information about general auditing techniques is presented in the CCPS book, *Guidelines for Auditing Process Safety Management Systems* (Reference 12-3) and Chapter 13 of the CCPS book, *Guidelines for Implementing Process Safety Management Systems* (Reference 12-4).

Typically, the MI program audit begins with a review of written programs/procedures and interviews with personnel involved to gain an understanding of the intended functions of the management system(s) associated with each audit topic. The auditor then determines if the management systems actually function as intended and if other requirements (e.g., jurisdictional requirements) are met. Table 12-1 summarizes an approach for evaluating MI management system functions.

When performing an audit of an MI program, the following issues should be considered.

***Level of Detail.*** Company health and safety audits will generally include the MI program; however, typical health and safety audit teams may not have the resources available or the expertise to thoroughly review MI. Some companies schedule specific MI program audits or even audits of specific MI practices (e.g., QA) to better focus on MI improvement.

***Objectivity.*** Some auditing can and should be done within a department to periodically verify program adherence (e.g., ITPM schedule adherence). However, periodic reviews by personnel from outside departments, sister plants, the parent company, and/or persons external to the company add perspective and can often elicit helpful suggestions for program enhancements.

***Documentation.*** Although not all program audits require documentation (unless the audit is for compliance with a jurisdictional regulation), most audits are documented with a report. Formal reporting helps ensure that any necessary

12. CONTINUOUS IMPROVEMENT OF MI PROGRAMS 235

TABLE 12-1
MI Audit Approach

| Audit Topic | Audit Approach Summary |
|---|---|
| MI Program Equipment | • Use facility tours and/or examination of piping and instrumentation diagrams (P&IDs) to select a representative sample of the different types of equipment items (e.g., pressure vessels, storage tanks, piping segments, pumps, PSVs, rupture disks, alarms and interlocks, release detectors, emergency shutdown [ESD] systems, and deluge systems) that are intended/required to be in the MI program, and confirm that each is actually in the MI program (i.e., MI program tasks are scheduled and records show that the tasks are being performed).<br>• At a minimum, MI programs should include all (1) equipment that under normal operating conditions contains a hazardous chemical, (2) equipment that is designed to relieve pressure or vent hazardous chemicals, (3) ESD systems, (4) safety devices/systems such as detection systems, suppression systems, alarms, and interlocks, and (5) critical utility systems. |
| ITPM Tasks and Schedules | • Review the ITPM tasks and schedules for the selected equipment to determine (1) if MI program tasks are identified and schedules exist for these tasks and (2) whether the tasks and schedules are appropriate.<br>• ITPM tasks and schedules should be (1) based on RAGAGEPs and/or manufacturer's recommendations and (2) more stringent and/or performed more frequently if dictated by facility experience (e.g., PSVs are on a more frequent ITPM schedule if they have failed prior tests/inspections).<br>• Review the completed MI program task records for the selected equipment, review/generate reports listing a backlog or past-due MI program tasks, and interview maintenance personnel to determine if (1) MI program tasks are being performed on schedule (as defined or understood within the facility), (2) tasks are being performed thoroughly (e.g., that all thickness measurement locations [TMLs] are addressed in each ultrasonic test), (3) the results of the MI program tasks are documented and the records contain all intended/required information, and (4) the results of MI program tasks are reviewed by a person who is qualified to determine if an equipment deficiency exists and recommend actions to take to correct the problem. |
| Written Procedures for MI Program Tasks | • Review the written MI program task procedures for the selected equipment and interview maintenance personnel as necessary to determine whether (1) written procedures exist for all tasks that could affect equipment integrity, (2) the content and level of detail in the procedures are sufficient based on the complexity and criticality of the tasks and the minimum skill level of the maintenance personnel, (3) the procedures are being kept current, and (4) the procedures are always accessible to personnel performing the tasks.<br>• To the extent that any written procedures provided in manufacturer/vendor manuals do not meet internal guidelines (e.g., address issues unique to the facility, such as personal protective equipment [PPE] and special tool requirements) and/or external requirements (e.g., regulatory requirements), such procedures must be supplemented to satisfy the need for written procedures. Note also that MI program procedures are typically kept current through periodic review by maintenance personnel and revision as necessary, and/or by including the procedures in the facility's management of change (MOC) process. |

## TABLE 12-1 (Continued)

| Audit Topic | Audit Approach Summary |
|---|---|
| MI Training | • Review a representative sample of MI training records, and interview a representative sample of maintenance personnel, to determine (1) which procedures and additional training topics specifically apply to which personnel (e.g., an overview of the processes and their hazards), (2) whether each craftsperson has been trained as intended/required on the procedures and topics that apply to their jobs before being required/allowed to perform a task, (3) whether training is also provided on an ongoing basis as new tools, equipment, techniques, and other changes are introduced, (4) whether personnel have been qualified, and certified as required by applicable RAGAGEPs, and (5) whether training is documented. Note that training must be documented for MI activities defined by RAGAGEPs requiring special certifications, such as American Petroleum Institute (API) inspection standards (e.g., API 510, API 653) and welding standards (e.g., ANSI/ASME B31.3).<br>• Note that training documentation may not be required by the MI program; however, if training is not documented, determine how the facility (1) internally tracks training to ensure that personnel receive needed training, (2) demonstrates to a compliance officer or auditor that the training has been provided, and (3) provides evidence that personnel understood the training. |
| Correcting Equipment Deficiencies | • Review any corrective action records (e.g., completed work orders), and interview maintenance personnel and/or operators to determine if corrective actions are consistently implemented before further use of the equipment, (2) adequate interim measures are consistently implemented to ensure safe operation when corrective actions aren't taken before further use of the equipment, and (3) equipment records document how and when identified deficiencies were corrected (and preferably any interim measures taken to ensure safe operation) and that any interim measures are tracked through permanent repair. |
| QA for Spare Parts/Equipment | • Review documentation (e.g., original equipment manufacturer [OEM] equipment lists, purchase orders, shipping papers, metallurgical analyses records), and interview personnel who order, receive, place into inventory, remove from inventory, and issue spare parts/equipment or materials/supplies (e.g., bolts, fittings, gaskets, o-rings, pump shafts, seals, welding rods) to determine whether (1) the correct items are consistently ordered, (2) any items received that are not what was ordered are consistently identified during the receiving process and returned to the supplier (or appropriately reviewed before being accepted), (3) received items are consistently placed correctly into inventory, and (4) the correct items are consistently removed from inventory for installation.<br>• Many facilities (1) order OEM spare parts/equipment/supplies whenever they are available, (2) use an MOC process when ordering other-than-OEM parts (even when the specifications appear to be the same), (3) require that persons ordering items personally examine them when received to better ensure that correct items are received, and (4) allow only designated personnel to place items into and remove items from inventory. Some facilities are also requiring suppliers to provide a metallurgical analysis or are testing metallurgy themselves when metallurgy is important. |

## TABLE 12-1 (Continued)

| Audit Topic | Audit Approach Summary |
|---|---|
| QA for Equipment Fabrication/Installation | • Interview personnel who order and/or oversee fabrication and installation of specialty equipment, and examine documentation as available (e.g., written QA plans, installation checklists) to determine whether the facility takes appropriate measures to ensure that equipment is fabricated and installed in a way that is suitable for the process application.<br>• Frequently, QA for equipment fabrication and installation involves (1) developing equipment fabrication specifications, (2) requiring various inspections and tests to be performed during fabrication to ensure that the equipment specifications are met, and (3) developing installation plans, schedules, and checklists to ensure that field installation issues are thoroughly and correctly addressed (e.g., bolt torquing, welding rod selection, gasket and packing materials, lubrication). |
| MI Documentation | • Determine whether all internally or externally required documentation of MI activities is created and retained.<br>• Some MI programs require only the following documentation:<br>  • Written ITPM task procedures<br>  • Required ITPM task lists and their schedules<br>  • ITPM task records showing the results of the tasks performed<br>• However, in most effective MI programs, the following additional documentation is also provided:<br>  • Roles and responsibilities assignments for MI program activities<br>  • MI program equipment list(s)<br>  • MI training records<br>  • Records showing that deficiencies have been corrected<br>  • Records (frequently in project files) of QA activities for equipment fabrication and installation<br>• Many facilities also describe the overall program in a written MI program document. |

follow-up actions are completed and provides an audit history for charting MI program success/progress. Consider documenting program practices in place, as well as program deficiencies.

***Audit Follow-up.*** When an audit is completed, facility management should promptly respond to audit findings and implement required changes. Management duties include (1) assigning follow-up responsibilities for every audit finding, (2) ensuring that a system is in place to track each audit finding to closure, and (3) documenting the closure of every audit finding. Management action (or inaction) is a key aspect in ensuring that the changes are implemented. Management should ensure that assigned personnel are (1) given the resources and support needed to implement the changes and (2) held accountable for implementing their assigned changes. Lack of diligence by management can result in a lost opportunity for improving the MI program.

## 12.3 PERFORMANCE MEASUREMENT AND MONITORING

Facilities are motivated to implement performance measurement systems for several reasons. Performance measurement enables facilities to:

- Monitor changes/improvements. As the MI program is improved or changed, the impact of changes/improvements can often be evaluated using performance measures.
- Make appropriate decisions for supporting the MI program. During the implementation and the life of the MI program, data can be used to make critical decisions. For example, decisions on staffing and budgetary issues for training and procedures can be supported via training and procedure measures.
- Track how the MI program is affecting safety and equipment reliability. Key benefits of an effective MI program are improvements in safety performance and increased equipment reliability. Safety and equipment reliability performance measures (e.g., near misses, equipment mean time between failure statistics) are often used to monitor the MI program. In fact, some organizations include these measures in the daily and weekly key performance indicators.
- Identify and publicize achievements. Measures can help show whether the allocated resources and the effort expended are truly contributing to the MI program, which can help individuals and groups understand and take pride in their contribution to the program.
- Ensure that the program is maintained. Performance that is measured tends to receive closer personal and organizational attention.

One approach for implementing a performance measurement system uses the following steps.

***Identify Appropriate Measures.*** The measures selected should be tied to the overall goals, objectives, and results of the MI program activities. A

## 12. CONTINUOUS IMPROVEMENT OF MI PROGRAMS

comprehensive system will include both lagging measures and leading indicator measures. The lagging measures directly measure the overall results of the MI program, usually in terms of the program goals and objectives (e.g., number of loss events, program costs). The leading indicator measures indirectly measure the overall MI program by measuring the program's activities (e.g., percent of ITPM tasks performed on time, average time equipment is operated in a deficient condition). Leading indicator performance measures help an organization predict the performance of the MI program. Because some direct measures of MI program performance are low-frequency events (e.g., number of loss events), leading indicators can be valuable for the decision-making process. In addition to identifying the measures, facilities need to determine the appropriate frequency for collecting and monitoring the performance measurement data. Many facilities collect and monitor MI program measures on a daily or weekly basis. Sometimes, these frequent measures are included in the facility's measurement "dash-board" (i.e., daily/weekly publication of the facility's key performance indicators).

***Collect and Analyze Performance Data.*** Many facilities with efficient and effective MI programs develop and implement systems for collecting and analyzing the data related to performance measures. Data collection systems can be paper-based and/or computer-based. Some of the computer-based data for the MI program can be generated from the computerized maintenance management system (CMMS) (e.g., overdue report for an ITPM task work order) or other databases associated with the CMMS. The data analysis includes any required calculations or data manipulation (e.g., filtering out of data anomalies) and tracking the results to look for trends, especially negative trends (e.g., increases in overdue ITPM task work orders). Facilities often use procedures or guidelines to ensure the consistency and quality of the data collected and the reporting of the analysis conclusions.

***Monitor the Performance Measures.*** Performance measurement systems should be set up to communicate necessary information to the personnel assigned to correct negative trends to improve the MI program. In addition, performance measures can be used to communicate the status of the MI program, and the improvements/effects resulting from the program, to the entire organization. Performance measures and the evaluation of the results should be periodically summarized and distributed to managers and decision makers within the organization. These reporting and communication activities help the MI program receive the continual attention that it needs.

In his book, *Risk-based Management: A Reliability-centered Approach*, Richard Jones presents Six Laws of Measurement (Reference 12-5):

1. *Anything Can Be Measured.* Computers have made it possible to measure almost everything.
2. *Just Because Something Can Be Measured Does Not Mean it Should Be.* As a result of Law 1, determining what should be measured becomes more important. Facilities should also clearly define how measurement results will be used (i.e., types of decisions to be influenced by the results).

3. *Every Measurement Process Contains Errors.* Even the most accurate measurement contains some error, and that error can adversely influence the result and thus the decision. Management needs to attempt to understand the magnitude of the error and the impact that errors can have on the results.
4. *Every Measurement Carries the Potential for Changing the System.* Measurement processes, especially those involving human intervention, can bias the measurement data and results. For example, simply measuring the air pressure in a tire results in a lowering of the pressure.
5. *The Human Is an Integral Part of the Measurement Process.* Humans involved in the measurement process affect and influence the measurements.
6. *(Jones' Law) You Are What You Measure.* Each measurement should be part of the overall management strategy for the MI program. The following guideline should be applied: each measurement must be directly related to achieving the mission of the organization (e.g., the MI program mission).

Developing a process map(s) of the MI program can help ensure that the correct measures are used to evaluate program performance (Reference 12-6). The process map should start with the program objectives (e.g., prevent catastrophic events) at the top and then work backwards through supporting program goals (e.g., preventing releases of hazardous materials) to MI program activities. Once the objectives, goals, and activities are identified, reasonable and practical measures can be developed for each of these items. While one or several measures could be identified for each item of the process map, an organization can realistically manage only a limited number of measurements. Figure 12-2 illustrates this process mapping concept and provides suggested measures.

Once facilities begin to collect the measurement data, the data should be converted into information that is used on a continual basis to assess and improve the MI program. Therefore, processes for analyzing and using the information should be defined and implemented, and management should commit to using the information to make decisions supporting the MI program.

## 12.4 EQUIPMENT FAILURE AND ROOT CAUSE ANALYSES

Facilities can incorporate lessons learned into the MI program by using a structured process to analyze equipment failures and associated management system failures. This process can include applying two different analysis techniques: failure analysis and RCA. While these two analysis techniques have many common aspects, their purposes, and thus their results, are different.

Failure analysis is designed to help personnel understand how the failure occurred and to determine and implement needed corrections and improvements to equipment. These corrections typically focus on redesigning the equipment item, altering its ITPM plan, and/or changing the process conditions. This analysis may

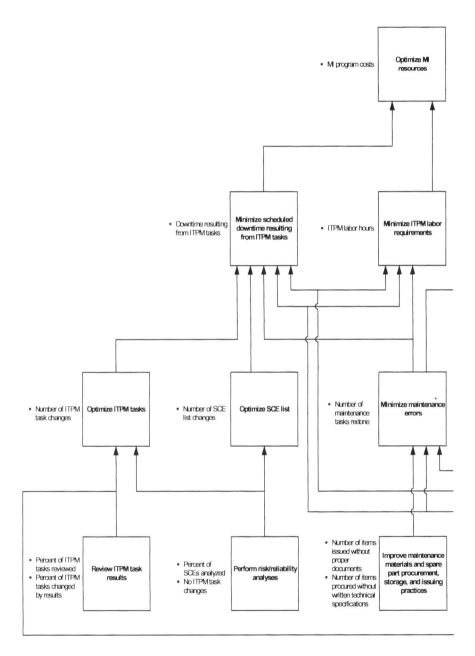

FIGURE 12-2 Example MI process map with suggested performance measures.

242 GUIDELINES FOR MECHANICAL INTEGRITY SYSTEMS

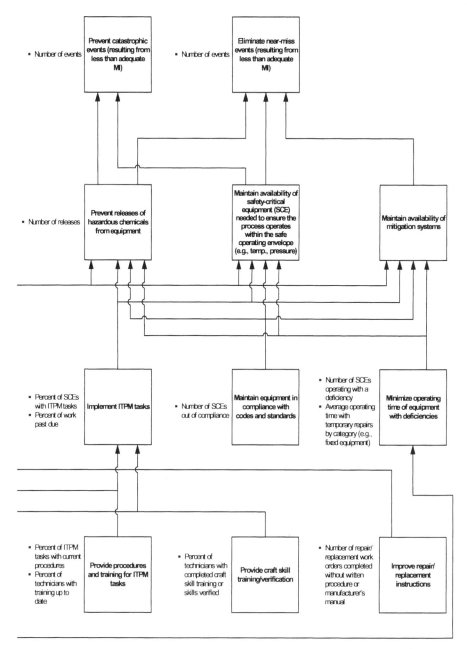

**Figure 12-2** *(Continued)*

## 12. CONTINUOUS IMPROVEMENT OF MI PROGRAMS 243

result in (1) changes that affect only the equipment item that failed or (2) changes that affect the failed equipment item as well as similar equipment items. Usually, this type of analysis is based on forensic engineering principles.

On the other hand, RCA is designed to discover why the failure occurred so that corrections and improvements can be made to the management systems that promote equipment integrity (e.g., MI program, operating procedures, engineering design practices, personnel training). Because RCA focuses on changes to management systems, the results tend to affect the equipment item that failed, similar equipment items, and seemingly unrelated equipment items that are managed by the same management systems. For example, if a pressure vessel leaks because of corrosion and RCA finds that nondestructive testing (NDT) activities were not being performed as planned, the RCA might result in improvements in the ITPM program that not only affect pressure vessel inspections, but also instrument calibration, rotating equipment lubrication, and so forth. The following subsections outline both failure analysis and RCA.

While the processes and results of failure analysis and RCA are distinctly different, they are quite often used together to analyze equipment failures. For example, a failure analysis provides vital technical information for the RCA. Similarly, the RCA process and associated tools can be used to provide useful information (e.g., potential damage mechanism for an equipment item, timeline of events) for the failure analysis.

### 12.4.1 Failure Analysis

Failure analysis processes focus on preserving, collecting, and analyzing evidence related to a failure. The analysis of the evidence begins with a macroscopic (i.e., broad) examination and progresses to a more microscopic (i.e., detailed) examination. Both nondestructive and destructive testing methods are sometimes used to determine the failure mechanisms that resulted in the failure. (See Appendix 4A in Chapter 4 for additional information on damage mechanisms.) Once the failure mechanisms are identified, the analysis team generates recommendations and corrective actions for reducing or eliminating the likelihood of a specific failure's recurrence. Figure 12-3 displays an eight-step analysis process (Reference 12-7). Each step is briefly described below:

*Step 1 – Evaluate Conditions at the Site.* Collect background information on the equipment and the process, as well as information related to the equipment conditions when the failure occurred (e.g., temperature, pressure, flow, operating mode).

*Step 2 – Perform Preliminary Component Assessment.* This typically involves performing a visual inspection. The visual inspection can uncover evidence of the type of failure that occurred.

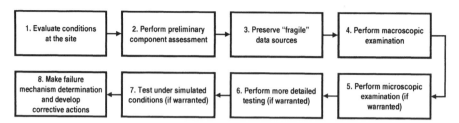

**FIGURE 12-3** Sample failure analysis process.

*Step 3 – Preserve "Fragile" Data Sources.* The evidence and data collected should be preserved to ensure that they are not lost and do not degrade.

*Step 4 – Perform Macroscopic Examination.* Perform (1) a visual inspection of component surfaces, dimensions, and so forth, with appropriate visual inspection tools (e.g., low power magnification, low power microscopes), (2) NDT (e.g., radiography, eddy current testing), (3) nominal chemistry testing (e.g., moisture content of gearbox oil), and/or (4) basic mechanical tests (e.g., hardness testing).

*Step 5 – Perform Microscopic Examination (if warranted).* For some failures, the preliminary visual inspection and macroscopic examination may not provide the information needed to determine the failure mechanism. In these cases, microscopic examination, using such tools as light microscopes, transmission electron microscopes, scanning electron microscopes, and energy-dispersive x-ray spectroscopy, may be needed to collect the needed information.

*Step 6 – Perform More Detailed Testing (if warranted).* Similar to Step 5, more detailed mechanical testing and/or chemical analysis may be needed to determine the failure mechanism. This testing typically focuses on determining the physical properties (e.g., metal hardness), compositions (e.g., chemical composition of lube oil), or other characteristics of material samples taken from the failed equipment. This testing many times results in destroying or altering the sample; therefore, personnel should verify that all visual examinations are completed before beginning such tests.

*Step 7 – Test Under Simulated Conditions (if warranted).* Experiments to reproduce a specific failure in a controlled environment (based on reasoned hypotheses from the available data) may provide verification of the hypothesis and insights into ways to prevent subsequent failures.

*Step 8 – Make the Failure Mechanism Determination and Develop Corrective Actions.* Based on the facts generated by the data analysis, conclude which failure mechanisms were significant contributors. Then identify specific events/ characteristics that caused the failure mechanisms to occur, and develop

## 12. CONTINUOUS IMPROVEMENT OF MI PROGRAMS

corrective actions for eliminating the causes and/or reducing the likelihood of the causes resulting in the failure.

The CD accompanying this book contains additional resources for performing equipment failure analyses (1) additional detailed information on the analysis steps, (2) an equipment failure analysis checklist, and (3) information on dominant failure mechanisms for typical plant equipment (e.g., tanks, vessels, pumps, piping). Chapter 8 of the CCPS book, *Guidelines for Investigating Chemical Process Incidents*, 2nd Edition, also contains information on equipment failure analysis. In addition, Richard Wulpi's book, *Understanding How Components Fail*, 2nd Edition, is a good reference on equipment failure analysis.

### 12.4.2 Root Cause Analysis

RCA is a process designed to investigate and categorize the root causes (i.e., equipment failures or human errors) of events with negative impacts (e.g., adverse safety, health, environmental, quality, reliability, and/or production effects). The term "event" is used to generically identify occurrences that produce, or have the potential to produce, these types of consequences. RCA is simply a tool designed to help identify not only WHAT and HOW an event occurred, but also WHY it happened. Only when investigators are able to determine why an event or failure occurred will they be able to specify workable corrective measures to prevent similar future events.

To effectively perform an RCA, a structured analysis process is needed. Many different RCA processes and tools are available. The following paragraphs discuss one approach, which includes the following four steps (Reference 12-8) (1) data collection, (2) data analysis, (3) root cause identification, and (4) recommendation generation and corrective action implementation. Each step is briefly described below:

*Step 1 – Data Collection.* The first step in the analysis is to gather four types of data (1) people (e.g., witnesses, participants), (2) physical (e.g., parts, chemical samples), (3) position (e.g., location of people and physical evidence), and (4) paper (e.g., procedures, computer data). Without complete information and an understanding of the event, the causal factors (CFs) and root causes associated with the event cannot be identified. The majority of time spent analyzing an event is spent gathering data. In addition, analysts may obtain information from other similar failures. This might include contacting personnel at other facilities that have experienced a similar failure and/or equipment vendor (or other outside) personnel who are knowledgeable about the failure.

*Step 2 – Data Analysis.* Organize the data and develop a model of how the event occurred. Two common methods for developing the model are CF charting and fault tree analysis (FTA). CF charting/FTA provides a structure for investigators to organize and analyze the information gathered during the investigation and to identify gaps and deficiencies in knowledge as the investigation progresses. The CF chart is simply a sequence diagram that describes the events leading up to an occurrence, as well as the conditions surrounding these events. FTA is a Boolean logic tool to help model the combinations of human errors, equipment failures, and external events that can produce the type of event being analyzed. Figures 12-4 and 12-5 (Reference 12-9) provide an example CF fault tree and CF chart, respectively.

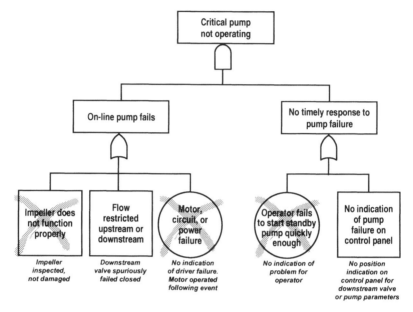

**FIGURE 12-4** Sample fault tree.

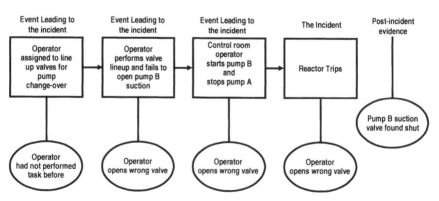

**FIGURE 12-5** Sample causal factor chart.

***Step 3 – Root Cause Identification.*** After all of the CFs have been identified, the investigators begin root cause identification. Many approaches use root cause diagrams, charts, or lists to aid the investigators in identifying the underlying

# 12. CONTINUOUS IMPROVEMENT OF MI PROGRAMS

reason(s) for each CF. (Examples of some of these root cause tools are provided on the CD accompanying this book.) Investigators use these tools to structure the reasoning process, which helps them answer questions about why particular CFs existed or occurred. The identification of root causes helps the investigator determine why the event occurred so that the problems surrounding the occurrence can be addressed. In addition, trending of the root causes of occurrences identified over a period of time can provide valuable insight concerning specific areas for improvement. As an added benefit of this process, RCA can be used to (1) help prevent the recurrence of specific events and (2) combine lessons learned from individual occurrences to identify major areas of weakness. This allows management to identify necessary actions to take before a seemingly unrelated accident or failure occurs.

***Step 4 – Recommendation Generation and Implementation.*** The next step is the generation of recommendations. Following identification of the root cause(s) for a particular CF, the analyst can generate achievable recommendations for preventing its recurrence. The root cause analyst is often not responsible for developing appropriate corrective actions based on the RCA recommendations. Therefore, analysis reports should provide enough information for those who are implementing the recommendations (or developing reasonable alternatives). Once the RCA team has developed its recommendations, facility management should evaluate the recommendations to determine the corrective actions to be implemented. In developing corrective actions, management must ensure that the corrective actions are practical and that they will be effective. In addition, management should communicate rejected recommendations, and the basis for the rejection, to the RCA team. Facilities should also ensure that a system is in place to track corrective actions to completion. When the entire occurrence has been charted, the investigators are better able to identify the major contributors to the incident. CFs are those contributors (human errors and/or component failures) that, if eliminated, would have either prevented the occurrence or reduced its severity. (Note: CFs are not the root causes of the event, rather, they are contributing causes to the event. In order to eliminate or reduce the likelihood of the event recurring, it is important that the root causes for CFs be identified and corrected.)

In many traditional analyses, the most visible CF is given all the attention. However, events are rarely caused by just one CF; they are often the result of a combination of contributors. Without identifying all of the CFs, the list of recommendations will likely not be complete. Consequently, the occurrence (and similar occurrences) may repeat because the organization did not learn all that it could from the original event.

The steps presented in this subsection outline one of many RCA processes that facilities can use to analyze equipment failures. For more information on RCA, see the CCPS book, *Guidelines for Investigating Chemical Process Incidents*, 2nd Edition.

## 12.5 REFERENCES

12-1. Occupational Safety and Health Administration, *Process Safety Management of Highly Hazardous Chemicals*, 29 CFR Part 1910, Section 119, Washington, DC, 1992.

12-2. ABSG Consulting Inc., *Reliability Management, Course 119*, Process Safety Institute, Houston, TX, 2004

12-3. American Institute of Chemical Engineers, *Guidelines for Auditing Process Safety Management Systems*, Center for Chemical Process Safety, New York, NY, 1993.

12-4. American Institute of Chemical Engineers, *Guidelines for Implementing Process Safety Management Systems*, Center for Chemical Process Safety, New York, NY, 1994.

12-5. Jones, R., *Risk-based Management, A Reliability-centered Approach*, Gulf Coast Publishing, Houston, TX, 1995.

12-6. ABSG Consulting Inc., *Enhancing Process Safety Performance, Course 134*, Process Safety Institute, Houston, TX, 2004.

12-7. Wulpi, D., *Understanding How Components Fail*, 2nd Edition, ASM International, Materials Park, OH, 1999.

12-8. ABS Group, Inc., *Root Cause Analysis Handbook, A Guide to Effective Incident Investigation*, Risk & Reliability Division, 1999.

12-9. American Institute of Chemical Engineers, *Guidelines for Investigating Chemical Process Incidents*, 2nd Edition, Center for Chemical Process Safety, New York, NY, 2003.

### Additional Source

Perry, Robert G. and J. Steven Arendt, *If You Can't Measure It, You Can't Control It; ProSmart Process Safety Management*, AIChE CCPS International Conference and Workshop, Toronto, Ontario, 2001.

# INDEX

Abbreviations, xxi-xxiii, 89
Abrasive wear, 36
Acceptable risk, 224-225
Acceptance criteria:
    characterized, 136
    defined, xxv
    installation, testing, and preventive maintenance, 33
    determination, roles and responsibilities, 56
        equipment-specific integrity management, 160, 163, 166, 169, 173, 177, 180
    risk, 210
Accident impacts, 201
Accountability, 10, 20
Acronyms, xxi-xxiii, 89
Activities backlog management, 16
Adhesive wear, 36
Administrative controls/procedures, 26, 79-81
Agitators, 19
Air systems, quality assurance, 107
Alarm systems, 21, 27-28, 32, 36, 157-158
Alterations, 96, 101-102
American Chemistry Council (ACC), 5
American Conference of Governmental Industrial Hygienists (ACGIH), 154
American National Standards Institute (ANSI):
    compressor RAGAGEPs, 148
    instrumentation systems, and automation, 204
    K61.1, 137, 142, 144
    process piping RAGAGEPs, 140

pump RAGAGEPs, 147
rotating equipment RAGAGEPs, 145
safety equipment standards, 158
S84.01 for safety instrumented systems (SISs), 20
American Petroleum Institute (API):
   certifications, 68
   compressor RAGAGEPs, 148
   emergency shutdown systems (EDSs), 169
   fans and gearboxes RAGAGEPs, 150
   fired heaters and furnaces RAGAGEPs, 152
   fixed equipment RAGAGEPs, 142
   inspection criteria, 160
   instrumentation and controls RAGAGEPs, 144, 146
   piping systems, 162-163
   pressure relief valves, 165-166
   pressure relieving devices, RAGAGEPs, 143
   pressure vessel RAGAGEP, 138, 191
   process piping RAGAGEPs, 140
   pump RAGAGEPs, 146
   Recommended Practice (RP) 580, 204, 218
   Recommended Practice (RP) 570, 142
   Recommended Practice (RP) 578, 113
   Recommended Practice (RP) 579, 131-133
   Recommended Practice (RP) 571, 36-37, 133
   Recommended Practice (RP) 572, Part II, 142
   rotating equipment RAGAGEPs, 145
   safety instrumental systems (SISs), 169
   Spec 12P, 141
      storage tanks, atmospheric and low-pressure RAGAGEPs, 139
   turbine RAGAGEPs, 149
American Society of Mechanical Engineers (ASME):
   Boiler and Pressure Vessel Code (BPVC), 131, 151
   B31.3, 190
   certifications, 68
   fired heaters/furnaces/boilers, 177
   fixed equipment RAGAGEPs, 38, 142
   instrumentation and controls RAGAGEPs, 144, 146
   pressure relief valves, 166-167
   pressure relieving devices RAGAGEPs, 143

# INDEX 251

pressure vessel RAGAGEP, 138
quality assurance (QA) standards, 110
Section IX welding, 163, 177
Section X, 141
    American Society of Nondestructive Testing (ASNT), certifications, 68
American Society of Testing and Materials (ASTM):
    D3299, 141
    D2583, 141
    D2563, 141
    E1476-97, Standard Guide for Metals Identification, Grade Verification, and Sorting Systems, 113
American Welding Society (AWS), certifications, 68
ANSI/ASME B31.3, 131, 142
Application of Safety Instrumented Systems for the Process Industries (ISA S84.01), 227-228
Approval, quality assurance procedures, 102
Area superintendent, roles and responsibilities, 12-13, 24, 55-57, 72-73, 93-94
As low as reasonably practicable (ALARP) risk matrix, 211
Atmospheric storage tanks, 18, 34, 101, 139
Audit(s):
    compliance with, 4
    components of, 231, 233-238
    departmental effectiveness, 14
    facility leadership role in, 14
    resolution recommendations and remediation, 16
    technical, 14
Authorization, quality assurance procedures, 102

Backup power supply/utility systems, 27, 37
Barricades, purpose of, 28
Barriers, purpose of, 28
Base Resource Document-Risk-based Inspection (API 581), 205, 218
Binning, 99
Blast walls, 28
Block diagram analysis, 108
Boilers:
    MI activity matrix, 158, 176-178
    state regulations, 17
Bonding system, 20, 28

Brittle fracture, 36
Buckling, 36
Budgeting and resources:
    initial implementation, 187-193
    ongoing efforts, 193-196
    program development, 183-187
Building ventilation, 21. See also Ventilation; Vent systems
Bunkers, purpose of, 28
Business plan, 9

Capital projects, 196
Catastrophic events, 20, 112
Causal factors (Cfs):
    defined, xxv
    root cause analysis, 245-247
Causal factor chart, xxv
Cautions, 78
CD, contents of, xvii-xix
Center for Chemical Process Safety (CCPS):
    as information resource, 78, 83, 87-89, 92, 97, 142, 144, 156, 205, 213, 226, 245, 247
    guidelines, see Guidelines, CCPS
    Technical Steering Committee, 1-2
Certification, xxv, 61, 63, 66, 68, 187, 194
Checklists:
    hazard evaluation, 97
    quality assurance, 109
    work completion, 116-117
Check sheets, 48
Chemical decompositions, 27
Chemical dosing pumps, 44
Chemical etches, 113
Chemical process industries (CPI), 1, 18, 155, 218
Chemical release mitigation system, 22
Chlorine Institute (CI), 137, 142, 144
Circuit diagrams, 22
Classroom training sessions, 65
Code of Federal Regulations (CFR), 1, 141
Common cause failure analysis, 109
Communications systems, 16, 28, 232

# INDEX
## 253

Compressed Gas Association, 142
Compressor(s):
    failure, 27
    frequency of ITPM tasks, 41
    RAGAGEPs, 148
    rebuilding, 37
Computer applications, <u>see</u> Computerized maintenance management systems (CMMS)
Computerized maintenance management system (CMMS), xxv, 42, 45-46, 48, 99, 191, 193, 197-198
Conceptual design phase, 107
Condition monitoring (CM):
    defined, xxv
    fitness for service (FFS) analysis, 132
    locations, 42
    techniques, 58-59
Construction/construction materials:
    quality assurance procedures, 96, 100-101, 108
    requirements, 15, 190
Construction supervisor, roles and responsibilities of, 106
Consultants, functions of, 184
Containment systems, equipment selection criteria, 20
Continuous improvement:
    failure analysis, 243-245
    learning from equipment failures, 231, 240-243
    model, 232
    overview of, 231-232
    program audits, 231, 233-238
        performance measurement system, establishment of, 231, 238-240
    root cause analysis, 245-247
Contractors:
    interface with mechanical integrity program, 4
    ITPM task selection team, 36
    MI procedure program, 93-94
    quality assurance, 99, 106, 116-117
    roles and responsibilities, 55-57, 129-130
    sample quality assurance plan, 116-117
    selection and management, 195
    training programs for, 61, 71, 191
Control mechanisms, facility leadership role in, 9-10

Control valves, equipment selection criteria, 28
Controls:
    administrative, 26, 79-81
    critical, 36
    documentation, 91
    environmental, 4
    frequency of ITPM tasks, 41
    instrumentation, 136, 144, 146
    inventory, 99
    purchasing, 115
    warehouse, 115
Conveyors, 19
Cooling systems, equipment selection criteria, 19, 27
Corrective actions, 16, 48, 50, 232
Corrosion:
    allowance, 33, 36
    engineers, ITPM task selection team, 33
    failure/damage mechanism, 36-37, 69
    monitoring, 58
    rate, 42, 50
    research studies, 221
Cost avoidance, defined, xxv
Coupon testing, 58
Cracking, failure/damage mechanism, 36-37
Craft training protocol, 82
Cranes, 28
Creep, 36
Critical controls, loss of functionality, 36
Critical instrument lists, 19
Critical operating parameters, xxv, 224

Damage/failure mechanisms, xxv, 36-37, 42
Damage Mechanism Affecting Fixed Equipment in the Refining Industry (API RP 571), 36-37, 133
Data collection systems, 239. See also Information gathering
Decision-making process, 9, 14
Decision trees, xxv
Decommissioning, 96, 103
Deficiencies, documentation of, 117. See also Equipment deficiencies; Equipment deficiency management

Deflagration and Detonation Arresters (CCPS), 142
Degradation rate confidence measures, 224
Design:
    detailed phase, 108-109
    equipment, 190
    new equipment, 135-136, 141, 159-181
    quality assurance requirements, 95-97, 107-109
    review, phases of, 107-109
    specifications, 32
Design of Industrial Ventilation Systems (ACGIH), 154
Detailed design phase, 108-109
Deterioration rate, expected, 42
Documentation:
    control systems, 91
    document management software, 199
    of effective training, 61, 64-66
    efficient ITPM program, 32
    equipment, 47
    equipment-specific integrity management, 136
    facility leadership roles/responsibilities, 10-13
    ITPM task results, 47-49
    MI activities, 161, 164, 167, 171, 175, 178, 181
    OEM, 236
    program audits, 237
        quality assurance procedures, 102, 110-111, 113-114, 116-117
    retention requirements, 49
    tracking, 98
Doors, blast-resistant, 28
DOT regulations, 141
Downtime, 18, 30, 201
Drawings:
    as-built, 32, 117, 190
    quality assurance procedures, 107
Ductile fracture, 36
Dump systems, 27
Dump tank drain valves, 26
Dust Explosions, Prevention and Protection (CCPS), 157
Dynamic monitoring, 58

Earthquakes, 21
Eductors, 18-19
EHS manager, roles and responsibilities of, 12-13, 93-94, 129-130
Electrical systems/equipment:
    equipment file information, 35
    grounding system, 20, 28
    quality assurance, 101, 107
    resistance testing, 20
    testing and monitoring, 58
Electric power supplies, 27. See also Backup power supplies
Electrochemical noise measurement, 58
Embrittlement, 36
Emergency communications systems, 20, 28
Emergency cooling systems, 27
Emergency inerting systems, 20
Emergency isolation valves, 44
Emergency planning and response, interface with mechanical integrity program, 4
*Emergency Relief System Design Using DIERs Technology* (CCPS), 142
Emergency response:
    equipment, 21, 28, 36, 157
    systems, characterized, 32
    training program, 63
Emergency scrubbers, 22
Emergency shutdown (ESD) systems, 25, 37, 144, 158, 168-171, 205
Emergency utility systems, 27
Emergency vents, 144
Employee(s), see specific job positions
    alarm systems, 157-158
    buy-in, 50, 82, 184
    participation in mechanical integrity program, 4
    training, see Training program
Engineers/engineering:
    data, 32
    on ITPM task selection team, 33
    maintenance, 55-57, 72-73, 93-94, 106, 129-130
    manager, 12-13, 93-94, 106, 129-130
    process, 55-57, 72-73, 93-94, 106, 129-130
    production, 55-57, 72-73, 93-94, 106
    project, 55-57, 93-94, 106, 129-130

quality assurance, 95-97
Environmental:
    controls, 4
    health, and safety recommendations, 37
    impacts, 201
    incidents, 18
    risk, 3
Environmentally critical equipment, 18
Equipment:
    availability, 45-46
    classes, xxv, 31-32
    condition trends, 15
    configuration, 42
    contractor-supplied, 96, 104
    deficiencies, see Equipment deficiencies
    design, 190
    documentation generation and maintenance, 56-57
    failure analysis, xxvi, 231, 240-243
    file information, 33-35, 103
    hazards, 112
    history records, 16
    incident investigation, 16
    integrity, 11, 14, 16
    life cycle of, 10
    list, program development resources, 185
    maintenance, responsibility for, 20
    manufacturers on ITPM task selection team, 36
    operating window, 14-15
    reliability, 4-5, 37
    repair, 16
    safe operating limits, 9-10
    selection factors, see Equipment selection
    used, 96, 103
    vendors, MI procedure program, 93-94
Equipment deficiencies:
    correction of, 236
    defined, xxv
    identification of, 50
    initial implementation resources, 187

management, see Equipment deficiency management
resolutions, 16
types of, 47-49
Equipment deficiency management:
acceptance criteria, 120-122, 124-125
communications, 127
equipment deficiency identification, 122-123
fitness for service (FFS) analysis, 131-133
initial implementation resources, 189, 193
permanent corrections, 127-128
process, 119-120
program development resources, 186
response to equipment deficiencies, 123, 126-127
roles and responsibilities, 128-130
Equipment selection:
criteria establishment, 17-21
defining level of detail, 17, 21-22
documentation, 17, 22-23
guidelines, sample, 25-28
program objectives review, 17-18
regulatory compliance, 17-18
roles and responsibilities, 23-24
significance of, 17
Equivalent pain matrices, 202
Erosion, 36
Essential Practices for Managing Chemical Reactivity Hazards (CCPS), 156
Evaluation Criteria for Reliability-centered Maintenance (RCM Standard JA1011), 215
Evaluation phase, in design review, 107
Event tree analysis, 225
Excess flow valves, 28
Eyewashes, 157-158

Fabrication:
characterized, 95, 98
management of, see Fabrication and equipment-specific integrity management
program audits, 237
quality assurance procedures/activities, 108, 195
Fabrication and equipment-specific integrity management:

characterized, 135-136, 141
MI activities, 159-181
Facility leadership roles/responsibilities:
auditing, 14
matrix, 10-13
organizational, 10
program documentation, 10
reporting mechanisms, 11, 14
Facility outages, scheduled, 46
Factory Mutual Research (FM), 142
Failure(s):
analysis, 243-245
equipment-specific integrity management, <u>see</u> Failures of interest and equipment-specific integrity management
mode, defined, xxvi
prediction of, 47
Failures of interest and equipment-specific integrity management:
characterized, 136
MI activities, 160, 163, 166, 169, 173-174, 177, 180
Fans, 19, 149
Fatigue, types of, 36
Fault tree analysis (FTA), xxvi, 42, 109, 225, 245-246
Ferrography, 58
Ferrous materials, 113
Filler metal composition, 113
Fire apparatus, 158
Fired heaters:
MI activity matrix, 158, 176-178
RAGAGEPs, 152
Firefighting equipment, 157-158
Fire mitigation systems, 22
Fire protection:
equipment, 20-21
insulation, 28
<u>Fire Protection Systems Inspection, Test and Maintenance</u> (NFPA), 153
Fitness for service (FFS):
analysis, 131-133
defined, xxvi
evaluation, 48, 50
Fixed process equipment, 37

Flame/detonation arresters, 142, 144
Flammable gas meters, 28
Flare systems, 22, 25-26, 28, 144
Flow alarms, 26
Fracture mechanics, 69
Freezing, in process equipment, 27
Fretting, 36
Functional tests, 19
Furnaces:
  MI activity matrix, 158, 176-178
  RAGAGEPs, 152

Gantt chart, xxvi, 183
Gears, RAGAGEPs, 150
General electrician, training matrix, 70
Geographical issues, 21
Glossary, xxv-xxviii
Grinding wheel spark test, 113
Guidelines, CCPS:
  Guidelines for Chemical Reactivity Evaluation and Application to Process Design, 156
  Guidelines for Engineering Design for Process Safety, 5
  Guidelines for Hazard Evaluation Procedures, Second Edition with Worked Examples, 97, 205, 213
  Guidelines for Implementing Process Safety Management Systems, 5
  Guidelines for Investigating Chemical Process Incidents, 2E, 245, 247
  Guidelines for Post-release Mitigation in the Chemical Process Industry, 156
  Guidelines for Process Safety in Batch Reaction Systems, 156
  Guidelines for Safe Automation of Chemical Processes, 144
  Guidelines for Safe Handling of Powders and Bulk Solids, 157
  Guidelines for Safe Storage and Handling of Reactive Materials, 156
  Guidelines for Writing Effective Operating and Maintenance Procedures,, 78, 83, 87-89, 92

Hatches, weighted, 18
Hazard:
  management system, 9-10, 15
  reviews, 97

Hazard and operability (HAZOP) analysis, xxvi
Hazardous material (HAZMAT):
    chemicals, 18
    loss of containment, 29
    training, 63
Heat:
    exchangers, equipment selection criteria, 28
    tracing, 28
High potential testing, 58
Hot spots, 28
Hot work permit, interface with mechanical integrity program, 4
Human error studies, 77-78
Human resource personnel, 72-73
Hurricane damage, 21
Hydrogen Institute (HI), 145

If/then statements, 88
Implementation resources, 187-193
Incident-free operations, 14
Incident investigation, 4
Independent protection layers (IPLs), 205
Industry practices, 37, 42
Inerting systems, 20, 28
Information gathering:
    ITPM task selection process, 32-33
    program implementation, 187-188, 190
    program procedures, 86
Inhibitor injection systems, 27
In-house history of equipment, 37
In-line filters, 18-19
In-service inspections, 133
Inspection(s), see Inspection, testing, and preventive maintenance (ITPM) program
    audio, 59
    codes, standards, and recommended practices, 37-38, 42
    contractors, ITPM task selection team, 36
    defined, xxvi
    documentation of, 23, 47
    equipment-specific integrity management, see Inspection and equipment-specific integrity management

fitness for service (FFS) analysis, 132
history, 32
importance of, 18
past-due, 16
personnel, see Inspection personnel
planning, 220
quality assurance procedures, 102
receipt, 99
reports, formal, 48-49
schedule, 16, 224
shop/site, 98
tasks, types of, 37
test equipment, 28
touch, 59
visual, 51, 59, 244

Inspection and equipment-specific integrity management:
characterized, 15, 135-136
MI activities, 159-181
roles and responsibilities of, 55-57

Inspection personnel:
ITPM task selection team, 33
managerial roles, 93-94, 106

Inspection, testing, and preventive maintenance (ITPM) program:
audits, 233-235
defined, xxvi
documentation, 57
equipment integrity issues, 30
equipment selection criteria, 16
history in equipment file information, 34-35
objectives of, 29
procedures, 79-81
program development resources, 185
quality assurance, 108
roles and responsibilities, 53, 55-57
task data collection and analysis software, 198-199
task execution and monitoring, 29, 46-52, 56
task implementation, 187-188, 191-194
task optimization, 194
task planning, see ITPM task planning

INDEX **263**

    task results documentation management, 57
    task selection, 55
    task work orders, 239
Inspectors, roles and responsibilities of, 93-94, 106, 129-130
Installation, see Installation and equipment-specific integrity management:
    program audits, 237
    quality assurance, 96, 100-103, 195
    temporary, 102-103
Installation and equipment-specific integrity management:
    characterized, 135-136
    MI activities, 159-181
Instrument/instrumentation:
    air supplies, 27
    characterized, 28
    and controls, RAGAGEPs, 146
    equipment file information, 35
    equipment selection criteria, 19-21, 27
    ITPM tasks, 41
    lists, 19, 23
    loops, equipment selection criteria, 22, 28
    quality assurance requirements, 101
Instrumentation, Systems, and Automation Society (ISA):
    emergency shutdown systems (ESDs), 169
    instrumentation RAGAGEPs, 144
    safety instrumented systems (SISs), 20, 169
Insulation, 19, 28
Insurance companies, 37
Integrity management, equipment-specific:
    electrical equipment, 136, 151
    fired equipment, 136, 151-153
    fire protection systems, 137, 153
    fixed equipment, 136-141
    instrumentation and controls, 136, 144, 146
    MI activity matrices, 158-182
    phases of, 135-137
    protective systems, 155-156
    relief and vent systems, 136, 142-144
    rotating equipment, 136, 145-150
    safety equipment, miscellaneous, 157-158

solids-handling systems, 156-157
utilities, safety-critical, 157
ventilation and purge systems, 154-155
Interlock instrumentation/systems, 28, 32, 36-37, 42
International Electrotechnical Commission (IEC):
emergency shutdown systems (ESDs), 169
functions of, 204
instrumentation and controls RAGAGEPs, 146
safety instrumental systems (SISs), 169
International Institute of Ammonia Refrigeration (IIAR), 137, 142, 144
International Organization for Standardization (ISO)-9000, 81
Inventory control, 99
ITPM task planning:
damage identification, 36
defined, 29
documentation, 32, 42
equipment failure mode, 36-38
inspection vs. replacement, 44-45
key attributes of, 30
operator performed tasks, 44
run-to-failure maintenance, 45
sample plan, in tabular format, 43
sampling criteria, development of, 42-44
task intervals, 36-42
task scheduling, 45-46
task selection, 30-38, 41-42
task selection decision matrix, 39-40

Jargon, use of, 89
Job classifications/subclassifications, 62-63
Job training protocol, 82
Jones' Law, 240
Jurisdiction requirements:
fabrication 98
risk-based inspections, 224
significance of, 32-33, 69, 71

"Kill" systems, 20
Kletz, Trevor, 194

INDEX **265**

Knowledge assessment, 61-64

Labeling, 99
Layer of protection analyses (LOPA):
    defined, xxvi
    facility safeguards, 20
    ITPM tasks, 32, 42
Layer of Protection Analysis, Simplified Process Risk Assessment (CCPS), 205, 226
Leadership, see Facility leadership; specific types of management
Leading indicator performance, 239
Leak repair clamps, 21
Life-cycle costs, 107
Lifting equipment, 28
Linear polarization, 58
Liquid:
    drain systems, 26
    expansion bottles, 28
    seals, 28
Litigation, 202
Localized corrosion, 36
Loss contributors, 108
Loss of containment, 36
Low-pressure storage tanks:
    characterized, 18
    quality of assurance requirements, 101
    RAGAGEPs for, 139
Lube oil systems, 21, 27
Lubrication fluid levels, 44

Macroscopic examination, failure analysis, 244
Magnetic particle testing, 47
Maintenance:
    contractors, ITPM task selection team, 36
    department, equipment selection roles, 23-24
    engineers, 33, 93-94, 106
    history, 32, 42
    ITPM task selection team, 33
    manager, 12-13, 72-73, 93-94, 106, 129-130
    material availability, 46

personnel, ITPM task selection team, 33
planner/scheduler, 55-57, 72-73, 93-94, 129-130
predictive, see Predictive maintenance
preventive, see Preventive maintenance
procedures, 79-81, 108
quality assurance procedures, 108
roles and responsibilities, 55-57
scheduling, 193
supervisors, 12-13, 72, 93-94, 106, 129-130
technicians, 93-94, 106, 129-130
Major industrial incidents, 77
Management of change (MOC):
    defined, xxvi
    implications of, 4, 79, 91, 224
    policy/program, 21, 23, 51-52, 63, 69, 102, 114, 193
Management responsibility:
    facility leadership roles and responsibilities, 9-14
    technical assurance responsibilities, 15-16
Management system, program development resources, 185
Manpower, maintenance scheduling, 191
Manufacturers:
    data sheets, 110
    functions of, 55-57
    quality assurance procedures, 108
    recommendations, 37, 42
Markov analysis, 225, 227
Markov modeling, 42
Material(s):
    composition analysis, 113
    contractor-supplied, 96, 104
    identification, 112-115
    quality assurance procedures, 98-99, 108, 112-115
    testing, 98-99
    tracking, 98
Material Verification Program for New and Existing Alloy Piping (API RP 578), 113
Mechanical integrity (MI) program:
    as company priority, 5
    defined, xxvi, 2
    development process, 3

INDEX     **267**

    expectations for, 3, 5
    objectives of, 19
    procedures, types of, 80
    relationship with other programs, 3-4
    success factors, 9
    synergies, 5
    training program, see Training program
Microscopic examination, failure analysis, 244
Mill testing, 112
Monomer vessels, polymer formation, 26

NACE International, certifications, 68
National Board of Boiler and Pressure Vessel Inspectors (NBBPVI):
    certifications, 68
    fired equipment RAGAGEPs, 151
    fixed equipment RAGAGEPs, 142
    standards, 137-138, 140, 143
National Electric Code (NFPA 70), 145, 151
National Fire Protection Association (NFPA):
    commonly used codes for fire protection systems, 153
    electrical code, 145, 151
    electrical system RAGAGEPs (NFPA 70), 151
    fired equipment RAGAGEPs, 151
    fixed equipment RAGAGEPs, 142
    instrumentation and controls RAGAGEPs, 144, 146
    protective systems, 155
    rotating equipment RAGAGEPs, 145
    safety equipment, RAGAGEPs, 158
    solids-handling systems, 156-157
    switch gears, 180
    ventilation and purge systems, 154
National Fuel Gas Code (NFPA 54), 155
New equipment design, equipment-specific integrity management:
    characterized, 135-136, 141
    MI activities, 159-181
Nitrogen:
    purge systems, 26
    storage, 41
Nondestructive testing (NDT):

characterized, 36-37, 42
defined, xxvi
risk management and, 243-244
techniques, 58-59, 116-117

Observation and surveillance CM techniques, 59
Occupational safety, significance of, 4, 18
Occupational Safety and Health Administration (OSHA):
    fixed equipment RAGAGEPs, 142
    process safety management (PSM) regulation, 1
    RAGAGEP requirements, 5
    safety equipment standards, 158
Oil analysis, 58
Ongoing training programs, 61, 69
Onsite storage, 20-21
On-the-job training, strength and weaknesses of, 65
Operations/operating:
    history, 42
    limits, 108
    procedures, generally, 4, 45, 79-80
    temporary procedures, 45
    types of procedures, 79-80
    window, xxvi, 14-15
Operational data, 32
Operations department/personnel:
    equipment selection roles, 23
    instrument lists, 19
    ITPM task selection team, 33
    roles and responsibilities of, 55-57, 93-94, 129-130
    training, 4
Optical emissions spectroscopy, 113
Organizational roles/responsibilities, 10
Original equipment manufacturer (OEM) manuals, 32, 45, 89-90
Outage windows, 51
Owner-user, defined, xxvi
Oxidizers, thermal, 28, 144
Oxygen, portable, 28

Pareto analysis, 107

Particle counter testing, 58
Parts retrieval, 100
Performance measurement system, xxvii, 231, 238-240
Performance monitoring, 59
Personal protective equipment (PPE), 79
Personnel, see Personnel qualifications and equipment-specific integrity management
    availability, 45-46
    safety, equipment selection criteria, 19
Personnel qualifications and equipment-specific integrity management:
    characterized, 136
    MI activities, 160, 163, 167, 170, 174, 177, 181
Pilot tests, 18, 189
Piping/piping systems:
    equipment file information, 34
    equipment selection criteria, 18-19, 21
    failure/damage mechanisms, 36
    MI activity matrix, 141, 158, 162-164
    pipe galleries, 28
    quality assurance requirements, 101
    task interval for inspection, 38
Piping and instrumentation diagram (P&ID), 22, 97
Pitting, 36
Plant evacuation alarms, 21
Plant manager, roles and responsibilities of, 55-57, 72-73, 93-94, 106, 129-130
Poison injection systems, 27
Polymer formation, 26-27
Positive material identification (PMI), xxvii, 95, 98, 114-115, 192
Power distribution equipment/systems, 19, 21
Power signature analysis, 58
Power supply systems, backup, 37
Pre-startup safety review (PSSR), 4, 101, 114
Predictive maintenance (PM):
    common, nondestructive testing techniques, 58-59
    defined, xxvii
    tasks, 37
Preliminary component assessment, 243
Preliminary design phase, 107-108
Premature failure, 37

Pressure alarms, 26
Pressure-boundary equipment, 36
Pressure equipment, quality assurance standards, 114-115
Pressure gauges, 44
Pressure relief valves (PSVs):
    characterized, 18, 28
    equipment selection criteria, 18, 22, 26-27
    MI activity matrix, 158, 165-168
Pressure safety valves, storage issues, 99
Pressure-vacuum valves, 18
Pressure valves, see Pressure relief valves (PSVs); Pressure safety valves
    ASME-codes, 38
    relief, see Pressure relief valves (PSVs)
    visual inspection of, 45
Pressure venting systems, 27
Pressure vessels:
    equipment file information, 34
    equipment selection criteria, 18-19, 21
    failure/damage mechanisms, 36
    MI activity matrix, 141, 158-161
    quality assurance requirements, 101
    state regulations, 17
    task interval for inspection, 38
Pre-startup safety review (PSSR), xxvii
Preventive maintenance (PM), see Preventive maintenance and equipment-specific integrity management:
    defined, 7
    quality assurance procedures, 103
    schedule adherence, 16
Preventive maintenance, equipment-specific integrity management:
    characterized, 135-136
    MI activities, 159-181
Proactive maintenance program, 30, 45
Probability of failure on demand (PFD), 225-227
Procedure requirements and equipment-specific integrity management:
    characterized, 136
    MI activities, 161, 164, 167, 170, 174, 178, 181
Procedures:
    adherence to, 15

INDEX **271**

    equipment-specific integrity management, <u>see</u> Procedure requirements and equipment-specific integrity management"
    program development resources, 185
    written, 98, 235
Process:
    downtime, reduction strategies, 30
    engineers/engineering, <u>see</u> Process engineers/engineering
    interlocks, 26
    motor failure, 26
    reliability initiatives, 5
    safety, 3-4, 18
    sewers, 28
    vessels, failure/damage mechanisms, 36
Process engineers/engineering:
    ITPM task selection team, 33
    manager, roles/responsibilities matrix, 12-13, 72
    roles and responsibilities of, 24, 93-94, 129-130
Process hazard analysis (PHA):
    characterized, 4, 32, 114, 157
    defined, xxvii
    equipment selection criteria, 20, 24
    implications of, 4, 114
    reports, 19
    in risk management, 225-226
    team, 19-20
Process map, 240-241
Process safety:
    coordinator, roles and responsibilities of, 55-57, 72-73, 93-94, 106, 129-130
    information, defined, xxvii
    management (PSM), equipment selection criteria, 17, 24
Procurement, 95, 97, 195
Production:
    department, equipment selection roles, 24
    engineers, roles and responsibilities of, 24, 93-94, 106
    outages, 202
    personnel, roles and responsibilities, 55-57
    supervisor, roles/responsibilities matrix, 12-13, 72, 93-94, 129-130
Product quality improvements, 18
Professional organizations, as information resource, 37

Program development, resources, 183-187
Program implementation:
　　budgeting and resources, 183-196
　　characterized, 16
　　return on investment (ROI), 200-202
　　software programs, 196-200
Program monitoring, 52
Program procedures:
　　accessibility of, 91
　　approval of, 87
　　benefits of, 77-78, 82
　　content, 78, 87-90
　　deficiencies in, 77-78
　　development process, 78, 81, 83, 85-88
　　drafts, 86
　　effective, 78
　　format, 78, 87-90
　　Guidelines for Writing Effective Operating and Maintenance Procedures (CCPS), 78, 83, 87-89, 92
　　hierarchy of, 81
　　implementation of, 78, 91-92
　　importance of, 77
　　maintenance of, 78, 91-92
　　needs identification, 78, 81-83
　　periodic review of, 87, 91
　　revisions, 87
　　risk-ranking, 83-84
　　roles and responsibilities, 78, 92-94
　　sources of, 78, 81, 90
　　types of procedures, 78-81
　　updating, 87
　　validation of, 86-87
Project(s):
　　cost monitoring, 183
　　　　engineers, roles and responsibilities of, 93-94, 106, 129-130
　　　　managers, roles and responsibilities of, 55-57, 93-94, 106
　　　　risk analysis, 107
Protective equipment, 32, 158. See also Personal protective equipment (PPE)
PSM Coordinator, roles/responsibilities matrix, 12-13
Public address systems, 28

Publicity, negative, 201-202
Pump(s):
    frequency of ITPM tasks, 41
    MI activity matrix, 158, 172-175
    RAGAGEPs, 147
    seal ruptures, equipment selection criteria, 26
    seals, characterized, 44
    vibration, 37
Purchasing:
    controls, quality assurance standards, 115
    personnel, roles and responsibilities of, 106
Purge systems, 28

Quality assurance (QA):
    alterations, 96, 101-102
    audits, 236-237
    characterized, 95
    construction, 96, 100-101, 112-115
    contractor-supplied equipment and materials, 96, 104
    decommissioning, 96, 103
    defined, xxvii
    design, 95-97, 107-109
    documentation, 48, 110-111
    engineering, 95-97
    fabrication, 95, 98
        Guidelines for Hazard Evaluation Procedures, Second Edition with Worked Examples (CCPS), 97
    importance of, 18
    initial implementation resources, 187, 189, 192
    installation, 96, 100-103
    ITPM task selection, 44
    life-cycle approach, 7
    material identification, 112-115
    predictive maintenance and NDT techniques, 58-59
    procedures, 79-80, 82
    procurement, 95, 97
    program development resources, 186
    program roles and responsibilities, 104, 106
    receiving, 95, 99
    repairs, 96, 101-103

rerating, 96, 101-102
retrieval, 95, 99-100
reuse, 96, 103
service contractor QA plan, sample, 116-117
spare parts, 96, 104, 108
storage, 95, 99-100
used equipment, 96, 103
vendor QA plan, sample, 110-111
verifications, 19
Quality control (QC), 95. See also Quality assurance (QA)
Quench systems, 27

Radiography, 47
RAGAGEPs:
    defined, xxvii, 5-6
    effective use of, 6, 49
    effect of, 5
    equipment acceptance criteria, 15
    equipment-specific integrity management, 137-140
    fitness for service (FFS) analysis, 133
    ITPM task planning, 44
    ITPM task selection process, 32-33, 41
    MI procedures, 82-83
    QA procedures, 113
    risk management techniques, 203-204
    safety equipment, 158-181
    training requirements, 69
Reaction:
    mitigation, 20
    quenching, 20
    thermal, 27
Reactor(s):
    block valves, 26
    upsets, equipment selection criteria, 27
Receiving, 95, 99, 195
Recognized and generally accepted good engineering practices, see RAGEGEPs
Recommended Practice for Electrical Equipment Maintenance (NFPA 70B), 145, 151

INDEX **275**

Recommended Practice for Emergency and Standby Power Systems for Industrial and Commercial Applications (IEEE 446), 151
Recommissioning procedures, 103
Recordkeeping, quality assurance issues, 97
Refresher training course, 61, 69
Refrigeration systems, 19
Regulatory:
    compliance, 17-18, 50, 202
    requirements, 32-33, 82-83
Relative ranking, 107-108
Release detectors, 20
Reliability:
    engineers, ITPM task selection team, 33
    improvements, 18
    management systems, 3
Reliability-centered maintenance (RCM) analyses:
    applications, 32, 41
    decision tree, 217
    defined, xxvii
    FMEA worksheet sample, 216
    rotating equipment, 145
Reliability-centered Maintenance and Reliability-centered Maintenance RCMI (Maubray), 205, 215
Relief catch pots, 22
Relief devices, equipment file information, 34
Relief dump tanks, 22
Relief system reviews, 97
Relief valves, see Pressure relief valves (PRVs)
    characterized, 27
    ITPM tasks, 41
    replacement of, 44
Remaining life:
    calculations, 47-48, 52, 132
    defined, xxvii
Remediation, 16
Repair(s), see Repair and equipment-specific integrity management
    in-service, 96, 101-102
    procedures, 80, 82
    temporary, 96,. 102
    types of, 96, 101-103, 132

Repair and equipment-specific integrity management:
    characterized, 136
    MI activities, 159-181 Replacement procedures, 80, 132
Reporting/reporting systems, see Documentation
    facility leadership roles, 10-11, 14
    project cost tracking, 183
Rerating, 96, 101-102, 132
Resistivity testing, 113
Resolution recommendations, 16, 128-130
Resource Guide for the Process Safety Code of Management Practices for Facilities, 5
Respiratory protection equipment, 158
Responsible Care®, 202
Retrieval, 95, 99-100
Return on investment (ROI):
    corporate reputation, reduced damage to, 202
    cost avoidance, 201-202
    importance of, 200
    improved equipment reliability, 200-201
    industry association commitments, 202
    liability, reduction of, 202
    regulatory compliance, 202
Reuse, 96, 103
Risk, defined, xxvii
Risk analysis:
    characterized, 32
    defined, xxvii
    recommendations, 37
Risk-based inspection (RBI):
    characterized, 69
    defined, xxvii
    equipment and process data, 219-221
    inspection planning, 220, 222-223
    inspection strategies, 220, 222-223
    management systems and tools, 220, 223-225
    overview of, 218-219
    program flowchart, 220
    research studies, 32, 41
    risk modeling, 220-222
Risk-based Inspection (API RP, 580), 205, 218

INDEX **277**

<u>Risk-based</u> <u>Management:</u> <u>A</u> <u>Reliability-centered</u> <u>Approach</u> (Jones), Six Laws of Measurement, 239-240
   Risk management, <u>see</u> Risk management program (RMP)
      common risk-based analytical techniques, 205-209
      decision process, 210-211
         failure modes and effects analysis (FMEA)/FMECA, 203, 206-209, 212-214
      fault tree analysis, 203, 206-209
         layer of protection analysis (LOPA), 203, 206-209, 225-228
      Markov analysis, 203, 206-209
         reliability-centered maintenance (RCM), 203, 206-209, 213, 215-217
      risk acceptance criteria, 210
      risk-based inspection (RBI), 203, 206-209, 218-225
      software, 199-200
   Risk management program (RMP), 17
   Risk modeling, 220
   Risk-ranking tool, 83
   Roles/responsibility matrix, <u>see</u> <u>specific</u> <u>job</u> <u>positions</u>
      equipment deficiency resolution, 128-130
      equipment selection criteria, 24
      ITPM task planning phase, 55
      ITPM task execution and monitoring phase, 56-57
      letter designations, 11
      quality assurance, 104, 106, 116
      significance of, 10
   Root cause analysis (RCA), xxviii, 243, 245-247
   Rotating equipment:
      equipment file information, 35
      ITPM tasks, 41
      selection criteria, 19, 21
   Routine inspections, 44
   Run-to-failure maintenance, 45
   Rupture disks, 18, 27-28, 144

<u>Safe</u> <u>Operation</u> <u>of</u> <u>Hydrofluoric</u> <u>Acid</u> <u>Alkylation</u> <u>Units</u> (API RP 751), 156
   Safety, Health, and Environmental (SHE) department, equipment selection roles, 23
      Safety and reliability analyses, 32
      Safety critical equipment, 18, 21
      Safety instrumented functions (SIFs), xxviii, 42, 205, 225

Safety Instrumented Functions (SIF)–Safety Integrated Level (SIL) Evaluation Techniques (ISA TR84.00.02-2002), 227
  Safety instrumented systems (SISs):
    defined, xxviii
    LOPA analysis, 225
    MI activity matrix, 144, 158, 168-171
  Safety integrity level (SIL), defined, xxviii
  Safety manager, roles and responsibilities of, 72-73, 106
  Safety Requirements for the Storage and Handling of Anhydrous Ammonia (ANSI-K61.1), 137, 142, 144, 156
  Safety showers, 157-158
  Safety systems, 36
  Sampling criteria, 47, 55
  Schedule(s), see Task schedule management
    adherence, 15
    extensions, 224
  Scrubbers, 28, 141
  Seal flush systems, 21
  Seal pots, 27
  Self-contained breathing apparatus (SCBA), 157-158
  Self-paced training, strengths and weaknesses of, 65
  Service history, in equipment file information, 34-35
  Sewer ventilation systems, 28
  Shock pulse analysis, 58
  Shutdowns, 42. See also Emergency shutdown (ESD) systems
  Signal drift, 36
  Site manager, roles/responsibilities matrix, 12-13
  Skills assessment, 61-64
  Society of Automotive Engineers, 215
  Software programs:
    computerized maintenance management systems (CMMS), xxv, 42, 45-46, 99, 191, 193, 197-198
    document management software, 199
    initial implementation resources, 187, 193
    ITPM task data collection and analysis software, 198-199
    program development resources, 186
    risk management software, 199-200, 223-224
    training program software, 199
  Solids processing equipment, 28
  Spare parts:

audits, 236
characterized, 45-46, 48, 96, 104, 108
quality assurance procedures/activities, 108, 195
storage of, 100

Spectrum analysis, 58

Spot trends, 47

Standard for Prevention of Fire and Dust Explosion for Manufacturing, Processing, and Handling of Combustible Particulate Solids (NFPA 654), 156

Standard for Purged and Pressurized Enclosures for Electrical Equipment (NFPA 496), 154

Standard on Explosion Prevention System (NFPA 69), 155, 157

Standard on Stored Electric Energy, Emergency, and Standby Power Systems (NFPA 111), 151

Startup criteria, 108

Steam systems, 19

Stewardship/status reports, 14

Storage, quality assurance procedures/activities, 95, 99-100, 195

Storage tanks:
    equipment selection criteria, 18
    explosive vapor space in, 20
    failure/damage mechanisms, 36
    fireproofing, 19
    inspection, 38
    internals, 28
    overfilled, 26

Storeroom personnel, roles and responsibilities, 55-57, 100, 106

Stress analysis, 69

Stress-corrosion cracking, 36

Structural components, 21

Structural steel, 28

Subject matter experts (SMEs):
    ITPM task selection team, 36
    program procedures, 86-87
    roles and responsibilities, 55-57

Supervisors:
    construction, 106
    maintenance, 24, 55-57, 72-73, 93-94, 106, 129-130
    production, 24, 55-57, 72-73, 93-94, 129-130
    responsibilities of, 15
    training programs, 63

Suppliers, selection criteria, 20. See also Vendors
Supply system, 41
Surveillance, visual, 44
Surveys, training, 75, 82
Switch gear, MI activity matrix, 158, 179-181
Synergies, significance of, 5
Systematic error, 227

Tanks, see Storage tanks
Task delays, 51
Task execution:
    acceptance criteria, 46-47
    documentation, 47-49
    implementation of, 49
    task results management, 49-51
    task schedule management, 51-52
Task frequency, 47, 52, 55
Task results management, 49-51
Task schedule management, 51-52, 55
Teamwork:
    ITPM task selection, 33, 36
    root cause analysis, 247
Technical assurance:
    acceptance criteria defined, 15
    defined, xxviii
    metrics establishment, 15-16
    technical content requirements, 15
    technical review, 16
Technical audits, periodic, 14
Technical evaluation condition, xxviii
Technical personnel/technicians:
    maintenance, 55-57, 72-73, 93-94, 106, 129-130
    roles and responsibilities of, 10
    training program, 61, 69
Temperature:
    indicating paint, 58
    measurement, 58
    monitors, 28
Testing, see Testing and equipment-specific integrity management:

INDEX                                                                           **281**

    equipment, 28
    independent service, 98
    quality assurance procedures, 102
    schedule adherence, 16
    tasks, 37
Testing and equipment-specific integrity management:
    characterized, 135-136, 141
    MI activities, 159-181
Thermal oxidizers, 28, 144
Thermal reactions, 27
Thermography, 58
Tracing systems, 27
Trade groups, as information resource, 37
Trade secrets, 4
Training manager, roles and responsibilities of, 72-73
Training program:
    approach considerations, 65
    audit of, 236
    certification, 61, 63, 66, 68, 187, 194
    contractor issues, 61, 71
    current workers, 61, 64
    development resources, 185
    effective, verification and documentation of, 61, 64-66
    flow chart, 62
    general electrician training matrix, 67
    instructors, 64, 69, 187-188, 190-191
    mechanical engineer, 70
    new workers, 61, 64
    ongoing, 61, 69
    outline of, 64
    refresher, 61, 69
    roles and responsibilities, 61, 71-73
    setting for, 64
    skills/knowledge assessment, 61-64
    software, 199
    program status, 15-16
    for technical personnel, 61, 69
    training survey, 75, 82
    written procedures, 98

workforce efficiency and, 201
Transformation, metallurgical, 36
Transportation equipment:
    integrity management, 141
    selection criteria, 20-21, 26
Turbines, RAGAGEPs, 149
Turnaround planning, 224

Ultrasonic testing, 47-48
Unavailability analysis, 227
Underwriters Laboratories Inc. (UL), 142
Uniform corrosion, 36
Uninterruptible power supplies (UPSs), 20, 28, 151
United States Coast Guard (USCG), 142
U.S. Environmental Protection Agency (EPA), 1, 5
Unit manager, roles and responsibilities of, 12-13, 55-57, 72-73, 93-94
Unit superintendent, roles and responsibilities of, 129-130
Used equipment, 96, 103
Utilities/utility systems:
    characterized, 20, 22, 32
    critical, 36
    equipment selection criteria, 26
    failures, equipment selection criteria, 27

Vacuum PRVs/conservation vents, 144
Vendor(s):
    equipment file information from, 34-35
    information supplied by, 190
    ITPM task selection team, 36
    MI procedure program, 93-94
    packages, equipment selection criteria, 21
    quality procedures, 97, 99, 108
    roles and responsibilities of, 129-130
    sample QA plan, 110-111
    selection and management, 20, 195
Vent handling systems, 28
Vent headers, 26, 144
Ventilation systems, equipment selection criteria, 18, 21-22, 27
Venturis, 18-19

# INDEX

Verification activity, xxviii

Walk about discussions, 14
Warehouse controls, quality assurance standards, 115
Warnings, 78, 88
Wear, failure/damage mechanisms, 36-37
Weather protection bags/socks, 28
<u>What</u> <u>Went</u> <u>Wrong: Case</u> <u>Histories</u> <u>of</u> <u>Process</u> <u>Plant</u> <u>Disasters</u> (Kletz), 194
What-if analysis, 107-109
Workmanship, quality assurance procedures, 102
Worksheets:
    FMEA sample, 214
    LOPA, 226-228
RCM FMEA sample, 216

X-ray spectroscopy, 113

# PUBLICATIONS AVAILABLE FROM THE CENTER FOR CHEMICAL PROCESS SAFETY OF THE AMERICAN INSTITUTE OF CHEMICAL ENGINEERS

Avoiding Static Ignition Hazards in Chemical Operations
Contractor and Client Relations to Assure Process Safety
Deflagration and Detonation Flame Arrestors
Electrostatic Ignitions of Fires and Explosions
Essential Practices for Managing Chemical Reactivity Hazards
G/L for Auditing Process Safety Management Systems
G/L for Analyzing & Managing Security Vulnerability of Fixed Chemical Sites
G/L for Process Safety in Batch Reaction Systems
G/L for Chemical Reactivity Evaluation and Application to Process Design
G/L for Chemical Process Quantitative Risk Analysis, 2nd Edition.
G/L for Design Solutions to Process Equipment Failures
G/L for Engineering Design for Process Safety
G/L for Evaluating Process Plant Buildings for External Fires and Explosions
G/L for Facility Siting and Layout
G/L for Fire Protection in Chemical, Petrochemical & Hydrocarbon Processing
G/L for Implementing Process Safety Management Systems
G/L for Investigating Chemical Process Incidents, 2nd Edition
G/L for Safe Process Operations and Maintenance
G/L for Technical Planning For On-Site Emergencies
G/L for Process Safety Documentation
G/L for Process Equipment Reliability Data
G/L for Process Safety Fundamentals in General Plant Operations
G/L for Safe Warehousing of Chemicals
G/L for Storage and Handling of Chemicals
G/L for Transportation Risk Analysis
G/L for Writing Effective Operating and Maintenance Procedures
G/L for Hazard Evaluation Procedures, 2nd Edition with Examples
G/L for Improving Plant Reliability through Data Collection and Analysis
G/L for Integrating Process Safety Mgmt, Environment, Safety, Health & Quality
G/L for Outsourced Manufacturing Operations
G/L for Pressure Relief and Effluent Handling Systems
G/L for Preventing Human Error in Process Safety
G/L for Safe Storage and Handling of Reactive Materials
G/L for Safe Automation of Chemical Processes
Inherently Safer Chemical Processes: A Life Cycle Approach
Layer of Protection Analysis: Simplified Process Risk Assessment
Practical Compliance With the EPA Risk Management Program
Revalidating Process Hazard Analyses
Tools for Making Acute Risk Decisions With Chemical Process Applications
Understanding Explosions
Wind Flow and Vapor Cloud Dispersion at Industrial and Urban Sites

CUSTOMER NOTE: IF THIS BOOK IS ACCOMPANIED BY SOFTWARE, PLEASE READ THE FOLLOWING BEFORE OPENING THE PACKAGE.

This software contains files to help you utilize the models described in the accompanying book. By opening the package, you are agreeing to be bound by the following agreement:

This software product is protected by copyright and all rights are reserved by the author and John Wiley & Sons, Inc. You are licensed to use this software on a single computer. Copying the software to another medium or format for use on a single computer does not violate the U.S. Copyright Law. Copying the software for any other purpose is a violation of the U.S. Copyright Law.

This software product is sold as is without warranty of any kind, either express or implied, including but not limited to the implied warranty of merchantability and fitness for a particular purpose. Neither Wiley nor its dealers or distributors assumes any liability of any alleged or actual damages arising from the use of or the inability to use this software. (Some states do not allow the exclusion of implied warranties, so the exclusion may not apply to you.)

WILEY